Performance-Based Contracts (PBC) for Improving Utilities Efficiency

Scientific and Technical Report Series No. 24

Performance-Based Contracts (PBC) for Improving Utilities Efficiency

Experiences and Perspectives

Jan Janssens, Didier Carron, Philippe Marin,
Philip Giantris and Tom Williams

IWA PUBLISHING

Published by **IWA Publishing**
Alliance House
12 Caxton Street
London SW1H 0QS, UK
Telephone: +44 (0)20 7654 5500
Fax: +44 (0)20 7654 5555
Email: publications@iwap.co.uk
Web: www.iwapublishing.com

First published 2018

British Library Cataloguing in Publication Data
A CIP catalogue record for this book is available from the British Library

ISBN: 9781780405957 (Paperback)
ISBN: 9781780405964 (eBook)
ISBN: 9781780409092 (ePub)

Contents

Principal Authors xiii

Contributing Authors xv

Preface xvii

Chapter 1
Performance-based contracts – setting the scene ***1***
Jan G. Janssens and Didier Carron

1.1 Introduction – PBCs and PPPs 1
  1.1.1 Specificity of PBCs 2
  1.1.2 Legal framework 5
  1.1.3 Designing a PBC 5
1.2 References 6

Chapter 2
PBC and results-based financing: The inverse approach ***7***
Didier Carron

2.1 Introduction 7
2.2 The Direct Problem 9
2.3 The Inverse Problem 11
2.4 Case Study 12
2.5 Conclusion 13
2.6 References 14

Chapter 3
PBC and energy efficiency ***15***
Didier Carron

3.1 Introduction 15

3.2 Energy Savings from Pumping ... 16
3.3 Other Energy Efficiency (EE) Possibilities ... 17
3.4 Performance-Based Contracts For Energy Efficiency ... 17
3.5 Designing and Implementing an EE Performance-Based Contract ... 19
3.6 Case Studies ... 19
3.6.1 Yerevan (Armenia) ... 19
3.6.2 Emfuleni (South Africa) ... 20
3.6.3 Larger contracts including an EE performance based component ... 21
3.6.4 Milwaukee ... 21
3.6.5 Lynn ... 22
3.7 References ... 22

Chapter 4
Internal performance contracts: A case of the National Water and Sewerage Corporation in Uganda ... **23**
Marine Colon and Silver Mugisha

4.1 Introduction: Why this Chapter? ... 23
4.2 IPCs at NWSC: Identity Card ... 24
4.3 Prerequisites for IPC Successful Implementation: Strong Foundations and Competition with the Private Sector ... 26
4.4 IPCs: An Integral Part of an Organisational Structure, A Management Style and an Organisational Culture ... 27
4.4.1 Key 1: Making managers accountable ... 27
4.4.2 Key 2: Motivate the staff ... 28
4.4.3 Key 3: Monitor and evaluate performance ... 28
4.4.4 Key 4: Manage resistance to change ... 29
4.5 IPCs: Only Constant is Change ... 30
4.5.1 Find the right balance between individual and collective incentive mechanism ... 30
4.5.2 Adjust the definition of performance over time ... 30
4.5.3 Find the right balance between control and confidence ... 30
4.5.4 Keep motivation high ... 30
4.6 Conclusion ... 31
4.7 Note ... 32
4.8 References ... 32

Chapter 5
Performance-based service contracts in Navi Mumbai ... **33**
Vijay Padmanabhan

5.1 Overview ... 33
5.1.1 Context ... 33
5.1.2 Current service delivery ... 34
5.1.3 Water supply management ... 35

5.2 The Performance-Based Contract Design ... 37
5.2.1 Overall nature of the PBC ... 37
5.2.2 Analysis of the contract structure ... 37
5.2.3 The procurement process ... 40
5.2.4 Performance monitoring ... 40
5.3 Implementation and Results Achieved ... 41
5.3.1 Results achieved ... 41
5.3.2 Sustainability ... 41
5.4 Lessons Drawn ... 42
5.5 References ... 43

Chapter 6
Financial comparison of NRW PBCs and conventional NRW projects ... ***45***
Alan Wyatt

6.1 Introduction ... 45
6.2 Methodology ... 45
6.3 Results ... 47
6.3.1 Project effectiveness ... 47
6.3.2 Project cost efficiency ... 48
6.3.3 Financial comparison ... 49
6.4 Conclusions ... 51

Chapter 7
Tegucigalpa PBC case study ... ***53***
David Michaud

7.1 Context: How The PBC Came About ... 53
7.2 The Design of The PBC ... 54
7.2.1 Overall nature of the PBC ... 54
7.2.2 Analysis of the contract structure ... 55
7.2.3 The procurement/tender process ... 59
7.3 Implementation and Results Achieved ... 59
7.4 Lessons Drawn ... 61
7.5 References ... 62

Chapter 8
Performance-based contracts – key design Issues ... ***63***
Victoria Delmon

8.1 Introduction ... 63
8.2 Activities where PBCs are Typically Used ... 64
8.3 External Factors that can Influence Design of the Contract ... 64
8.4 Parties to the Contract ... 65
8.5 Scope of the Contract and PBC Contractor Obligations ... 66

8.6 Obligations of the Contracting Authority 69
8.6.1 Monitoring and supervision 70
8.6.2 Duration and sustainability of contract 70
8.6.3 Fees and incentives 71
8.6.4 Dispute resolution 71
8.6.5 Early termination 72
8.6.6 Procurement 72
8.7 Acknowledgements 73
8.8 Other Resources 73

Chapter 9
NRW reduction optimization framework* *75
David Ehrhardt

9.1 Introduction 75
9.2 Optimizing Physical Losses 76
9.2.1 The point at which physical losses are optimized 76
9.2.2 Deriving the supply curve for physical loss reduction 77
9.2.3 The various benefits of physical loss reduction 78
9.2.4 Estimating the value of deferred capital costs 79
9.3 Optimizing for Service 81
9.3.1 The interaction between physical losses and hours of supply 81
9.3.2 Valuing benefits from increased duration of supply and increased access 83
9.4 Optimizing Commercial Losses 84
9.5 A General Optimization Framework 86
9.6 Conclusions 87
9.7 Note 88
9.8 References 88

Chapter 10
How to improve water services performance* *89
Brice Cabibel, Fabien Mainguy and Florie Lozivit

10.1 Introduction 89
10.2 A Strong Water and Sanitation Governance 90
10.3 A Smart PBC Design 91
10.3.1 Baselines, targets and incentives 92
10.3.2 Ensure mitigation of risks and leave room for flexibility 93
10.3.3 Develop an integrated approach 95
10.4 A Resilient Economic Model 97
10.4.1 Enhanced financial sustainability 97
10.4.2 A sustainable territorial anchorage 98
10.5 Conclusion 100
10.6 References 101

Chapter 11
Performance-based contracts (PBC) and regulatory issues ***103***
Sanford V. Berg and Robert N. Gakubia

11.1 Background 103
11.2 Monitoring and Evaluating Utilities 104
11.2.1 Building information systems with key performance indicators (KPIs) that enable regulators to monitor, and evaluate operators 104
11.2.2 Setting performance targets and monitoring performance with limited data 106
11.2.3 Developing incentives for quality improvements, cost containment and network/service expansion 106

Chapter 12
Peer-to-peer partnerships operational for sustainable water services ***109***
Siemen Veenstra

12.1 Introduction 109
12.2 The Water Operator Partnership 111
12.3 Components of an Operational Partnership 112
12.3.1 Preparing a WOP 113
12.3.2 Incentives: yes or no? 113
12.3.3 Priority setting in performance improvement 114
12.3.4 Conditional factors for WOP success 114
12.4 Summary and Recommendations 118
12.5 References 119

Chapter 13
Performance-based contracts in Malawi: Teamwork works ***121***
Siemen Veenstra, Alfonso Chikuni and Andrew Thawe

13.1 The Context: How the PBC Came About 121
13.2 The Design of the PBC 123
13.2.1 Overall nature of the PBC 123
13.2.2 What is the PBC about? 123
13.3 Performance Contract Design 124
13.3.1 Services to be provided 125
13.3.2 The remuneration structure 125
13.3.3 Performance regime: a “stick and carrot” approach 125
13.4 The Procurement/Tender Process 127
13.4.1 The pre-qualification stage 128
13.4.2 The proposal stage 128
13.5 Implementation and Results Achieved 129
13.5.1 The “office” phase 129
13.5.2 The “operations” phase 129
13.5.3 The extension phase 131
13.5.4 Remuneration 131

13.6 Did the Contract Achieve and Sustain Results? . . . 132
13.6.1 Internalities . . . 132
13.6.2 Externalities . . . 132
13.7 Reflections on Performance-Based Service Contracting . . . 133
13.8 Acknowledgements . . . 134
13.9 References . . . 134

Chapter 14
Performance-based affermage contracts . . . ***135***
Vivian Castro-Woolridge, Didier Carron and Jan Janssens

14.1 The Basics of the Affermage Contract . . . 135
14.2 Differences and Similarities Between the Affermage and Other Operations Contracts . . . 136
14.3 Affermage Contracts from the 19th Century Until Today In France . . . 136
14.4 The Way to a Second Generation of Affermage Contracts . . . 137
14.5 The Performance-Based Affermage (PBA) Model . . . 138
14.6 The Earning Logic . . . 138
14.7 Technical Performance Indicators . . . 139
14.8 Financial Performance Indicators . . . 140
14.9 Case Study: The Sedif . . . 140
14.10 Sedif: Return of Experience 2011–2013 . . . 142
14.11 Conclusion . . . 143
14.12 References . . . 144

Chapter 15
Performance-based contracts, the Aroona Integrated Alliance experience . . . ***145***
Jérôme Douziech and Donna Buchanan

15.1 Case Study Detail . . . 145
15.1.1 The Aroona Integrated Alliance to deliver improvements in water and wastewater services . . . 145
15.1.2 A commercial framework set up to generate the right behaviour . . . 146
15.1.3 Alliance governance . . . 147
15.1.4 The "good to great" strategy to deliver a step change . . . 148
15.1.5 A governance model is in place to ensure business improvements are sustainable . . . 149
15.1.6 A Business improvement program designed to deliver sustainable cost efficiencies . . . 149
15.1.7 The cultural and the change management challenge . . . 152
15.1.8 Main achievements two years after the Good to Great strategy inception 153
15.2 Conclusion . . . 153

Chapter 16
Experience from Eastern Europe ***155***
Augustin Boer, Teodor Popa and Dobromir Simidchiev

Chapter 17
Performance-based contracts (PBCs) for improving utilities efficiency: Experiences and perspectives: The way forward and perspectives/trends . . . ***163***
Didier Carron and Jan Janssens

17.1 Summary of Discussions of 31 October 2016 163

Index **165**

Principal Authors

Didier Carron
Naldeo SA

Jan G. Janssens
JJC sàrl

Philip Giantris
Former President of Valu Add Management Services

Philippe Marin

Tom Williams
IWA

Contributing Authors

Alan Wyatt
Independent Consultant

Siemen Veenstra
Vitens Evides International
Zwolle, The Netherlands

Andrew Thawe
Blantyre Water Board, Malawi
(he is the former CEO BWB)

Dobromir Simidchiev
Chief Executive Officer (CEO)
Hydrolia AD, Bulgaria
dsimidchiev@hydrolia.bg

Teodor Popa
Economic Director (CFO)
Compania Apa Brasov S.A.
Romania
dorupopa@apabrasov.ro

Vijay Padmanabhan

Silver Mugisha
NWSC

David Michaud
The World Bank, Water

Florie Lozivit
International Water Projects Division of SUEZ.

Fabien Mainguy
International Water Projects Division of SUEZ.

Sanford V. Berg
Public Utilities Research Commission (PURC)

Augustin Boer
Senior Consultant
BDO Business Advisory S.R.L
Romania
augustin.boer@bdo.ro

Donna Buchanan
Aroona Alliance, Perth, WA, Australia

Brice Cabibel
International Water Projects Division of SUEZ.

Vivian Castro-Woolridge

Alfonso Chikuni
Lilongwe Water Board, Lilongwe, Malawi

Marine Colon
AgroParisTech, G-EAU, Montpellier
AgroParisTech, Univ Montpellier, Irstea, CIRAD
IRD, Montpellier SupAgro

Victoria Delmon

Jérôme Douziech
Aroona Alliance
Water Corporation
Australia

David Ehrhardt

Robert N. Gakubia
Water Services Regulatory Board (WASREB)

Preface

The private sector can help to boost water utilities towards improving efficiency and technology within short periods of time. However, involving the private sector for water management is a very sensitive issue in all countries, both politically and socially. There is by far no single answer to this issue since it depends on a wide range of factors.

Between normal procurement, which is necessarily input based, and concessions, public-private partnership (PPP) or build-operate-transfer (BOT) contracts (which are highly engaging for the public party), performance-based contracts (PBCs) are a very valid and effective option to take advantage of private expertise and capacity to commit under the control of the public authority. Performance-based contracts are focused, limited in time and their financial logic is virtuous.

Even if PBCs have been used so far mainly for non-revenue water (NRW) reduction, with much success even if the way to success is not always straightforward, PBCs can be used to tackle many other issues such as energy saving, renewable energy production from sludge, smart metering, efficient management through management contracts, etc. However, there are no two similar cases when using PBCs. This is due, on the one hand, to the variety of situations, technical but also historical and political, and, on the other hand, the variety of contractual options, especially regarding how to express the performance and then the remuneration of the contractor.

A PBC today will still be tailor-made for each case considered, which makes the task of designing and implementing a PBC much more difficult than just applying a formula, but also much more attractive and hopefully successful.

Consequently, this book is not a course on PBCs, which would be a very ambitious project. Instead, it is a collection of case studies which reflect experiences that can also serve as a useful background when projecting a PBC.

This book is a collective work, and contributors are here warmly thanked for their contributions. It helps to reaffirm the feeling that the water community is a family, and we want to express our sincere thanks to the IWA, who actively contributes to creating the tools and opportunities to enable the wider water community to share their experience from a bottom-up approach, and therefore continuously build a collective knowledge base.

Jan Janssens and Didier Carron

Chapter 1

Performance-based contracts – setting the scene

Jan G. Janssens and Didier Carron

Performance-based contracts (PBCs) are a special case of public–private partnership (PPP) contracts and they widely differ from traditional public sector, input-based, contracting methods. The whole idea of PBCs is to make contracts more efficient, by optimizing incentives and allocating risks to the party more able to bear or mitigate it, so that better results can be expected. Designing PBCs is however less simple than designing conventional contracts, as it involves setting objectives, defining key performance indicators (KPIs) reflecting these objectives and selecting a remuneration formula based on these KPIs so as to become a win–win agreement for both the utility and its contractor. Such contracts should also comply with the existing legislations on public contracting.

1.1 INTRODUCTION – PBCs AND PPPs

While traditional input-based contracts often bring disappointing results, delegating operations, or part of operations, to private management under a performance-based public–private partnership (PPP) offers a valuable solution, more efficient than traditional input-based contracts.

The whole idea of performance-based contracting (PBC) is to make contracts more efficient, by optimizing incentives and allocating risks to the party more able to bear or mitigate it. Properly designed, a performance-based contract can achieve better results than are normally achieved using traditional public sector, input-based, contracting methods. The advantage comes from a better alignment of incentives with output objectives, coupled with more contractor flexibility in design and implementation to facilitate innovation and encourage efficiency.

Performance-based contracts belong to two categories of contracts:

- *Results-based financing contracts*
  Results-based financing (RBF) is an umbrella term linking payment to achieving a predetermined performance target.

Authors: Jan Janssens, Didier Carron, Philippe Marin, Philip Giantris and Tom Williams
doi: 10.2166/9781780405964_0001

RBFs have become common practice in highways and roads, healthcare and energy sectors.

In the water and sanitation sector, RBFs are still less well spread; the most common application is through output based aid (OBA), often and mainly targeted at increasing access to the urban poor.

Performance based contracting is an RBF where the remuneration of the contractor is directly related with its performance through quantified key performance indicators (KPIs).

- *Public–private partnership contracts*
  PPP involves a contract between a public sector authority and a private party, in which the private party provides a public service or project and assumes substantial financial, technical and operational risk in the project.

In some types of PPP, the cost of using the service is borne exclusively by the users of the service and not by the taxpayer. In other types (notably the private finance initiative), capital investment is made by the private sector on the basis of a contract with government to provide agreed services and the cost of providing the service is borne wholly or in part by the government. Government contributions to a PPP may also be in kind (notably the transfer of existing assets).

There are usually two fundamental drivers for PPPs:

- First, PPPs are claimed to enable the public sector to harness the expertise and efficiencies that the private sector can bring to the delivery of certain facilities and services traditionally procured and delivered by the public sector.
- Second, a PPP is structured so that the public sector body seeking to make a capital investment does not incur any borrowing. Rather, the PPP borrowing is incurred by the private sector vehicle implementing the project. Generally, financing costs will be higher for a PPP than for traditional public financing, because of the private sector higher cost of capital. However, extra financing costs can be offset by private sector efficiency, savings resulting from a holistic approach to delivering the project or service, and from the better risk allocation in the long run.

By comparing both definitions, one can see that:

- PBCs are PPPs in the sense that PBCs clearly involve risk taken by the contractor, which is a major differentiation from a traditional service or works contract.
- However, as PBCs typically do not include much investment, PBCs are thus "light" PPPs which, for implementation, do not need all the tools used for typical PPPs such as special purpose vehicles (SPVs), complex guarantee schemes or complex financing schemes.

1.1.1 Specificity of PBCs

It can be seen that PPPs, which are mainly *service contracts*, can be designed as performance-based contracts.

The design of a PBC includes thus three specific components which are not so much used in common PPPs:

- Defining objectives. These objectives should be both ambitious and realistic;
- Designing key performance indicators (KPIs) reflecting these objectives. The design of these KPIs is crucial as they will drive the remuneration of the contractor, and can be quite tricky;
- Correlatively finding the right balance between necessary empowerment of the contractor and control by the utility.

The implementation of a PBC also has much specificity, such as:

- The need to establish a validated baseline. This is an absolute precondition for an effective incentive structure with realistic time-bound performance targets – inaccuracy of baseline is an issue and a constraint.
- The overall target of improving performance and building capacity.

However there also exists some specific ways to develop PBCs, especially third party financing. Third-party financing (TPF) means that project financing comes from a third party, e.g. a finance institution, and not from internal funds of the contractor or of the customer. The finance institution may either assume the rights to the energy savings or may take a security interest in the project equipment.

This type of mechanism is quite often used for performance based energy efficiency contracts (PBEECs), where the contractor is an energy savings company (ESCO).

As described by the EU Commission, there are two conceptually different TPF arrangements associated with EPC; the key difference between them is which party borrows the money: the ESCO or the client.

- The first option is that the ESCO borrows the financial sources necessary for project implementation.
- The second option is that the energy user/customer takes a loan from a finance institution, backed by an energy savings guarantee agreement by the ESCO. The purpose of the savings guarantee is to demonstrate to the bank that the project for which the customer borrows will generate a positive cash flow, i.e. that the savings achieved will certainly cover the debt repayment. Thus, the energy savings guarantee reduces the risk perception of the bank, which has implications for the interest rates at which financing is acquired. The "cost of borrowing" is strongly influenced by the size and credit history of the borrower.

At a first glance, there are no good reasons why PBCs are so little favored, as PBCs have the potential to be more efficient contracts than the traditional, input-based contracts. Despite their potential, PBCs are not necessarily always the most appropriate mechanism to meet a utility's objectives in a given situation. The following criteria will help determine whether the utility's conditions are conducive to a successful PBC:

- Do potential benefits outweigh the costs? The utility needs to do a cost–benefit analysis, pondering especially whether there is real value in providing the contractor with flexibility on some tasks, and if this compensates for the higher price which the contractor shall probably require in exchange for taking more risks;
- Are quality baseline data available? This is essential for being able to properly measure performance improvements (comparing "before" and "after"). When no reliable baseline is in place, this may make the contract risky due to potential conflicts during implementation;
- Can the PBC be ring-fenced from utility management? To justify a bonus payment, performance improvement needs to be clearly linked to the specific actions of the contractor;
- Do public procurement rules allow for contracting services under a PBC? The potential constraints and limitations of the legal framework need to be well understood. As PBCs are not conventional procurement contracts and are also not conventional PPP contracts, as they typically do not involve strong investments by the private partner, they might be not defined by the law and thus might not be authorized.

  Whether PBCs may thus seem an "obvious" way of procurement to ensure results, it is also essential to say they are quite complex to implement and then to manage:

- Defining ambitious but realistic objectives through performance indicators with quantified targets is not an easy task. This includes some kind of forecasting, which is never an easy job.

- Finding the appropriate risk level for the contractor is not easy, and is precarious for the contracting utility: in case too high a risk is allocated to the private party, there is a high probability that the tender will be not responsive, or, of conflicts arising during execution of the contract as a result of non-reached targets; in case too low a risk is allocated to the private party, then too high a profit may be achieved.
- The achieved performance needs to be measured and this is crucial as it determines the profit of the contractor; this is often a delicate process due usually to unexpected external events occurring during the life of the contract.
- The contractor should be empowered, which might create conflicts with the existing organization.

A major shift in PPP contracting in recent years has been the extensive use of KPIs for a tighter and more transparent control by the public entities. While private financing can be one of the main attractions of PPPs, it is the efficiency gains and improved service quality and compliance brought by the private sector's management systems and innovative technologies and techniques that make PPPs an attractive mode of delivery in the water and sanitation sector. Efficiency gains are mostly captured by reduction in non-revenue water (NRW), improvement in billing and bill collection and labor productivity. Performance basis is a feature that applies to all types of PPP contracts and services and can be mainstreamed, in public–private and in public–public partnerships (see Figure 1.1). This requires setting up enabling conditions, to be implemented through governments/municipalities, and utilities/operators, with the support of international financing institutions (IFIs).

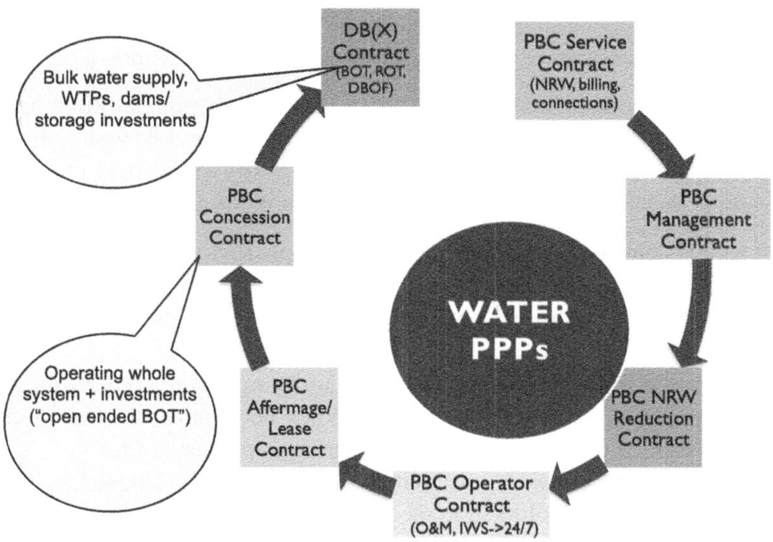

Figure 1.1 PBC Features can be applied to a wide variety of PPP contracts.
Adapted from V. Delmon, World Bank, 2015.

Another challenge is how to make sustainable the gains of efficiency, reached through PBCs. This may require a balance be established between improving performance and building capacity throughout the duration of the contract.

It thus can be seen that through more sharing of know-how and successful case studies, and by promoting PBCs as an essential way of procurement, PBCs should significantly develop in the water sector in the future, with application to new challenges of water utilities such as energy efficiency or smart metering.

In this context, this book aims at presenting the wide spectrum of the use of PBCs in the water sector, covering a large variety of applications, presenting their effectiveness but also the key success factors with respect to the constraints listed above, and showing the potential promising future uses of PBCs.

1.1.2 Legal framework

In many countries the legal framework distinguishes two or three separate lines of procurement:

- "Standard" procurement for goods and services
- Concessions
- PPPs

A few countries have specific legislation on PBC.

Typically, "standard" procurement may not include performance-based contracts, and limits the possibilities of result based contracts to a few incentives and penalties. This might force public buyers to use PPP procedures to conclude a performance based contract. However, these procedures are often quite heavy, especially regarding compulsory preliminary assessment and/or administrative authorizations, limitation of scope of PPP contracts, etc. As explained above, these quite complex procedures might be too heavy for a "simple" PBC contract.

Laws on concession typically allow performance based remuneration of the contractor, however they might be restricted to contracts including operation duties, which might not be the case for a "PBC service contract". So, as can be seen, legislation might be a strong obstacle to public entities, either to select such an option for procurement or for the procurement process itself.

1.1.3 Designing a PBC

A well-designed PBC is a win–win for both the utility and its contractor.

- While the contractor is required to take more risks, he has also more incentives to deliver tangible results; but in setting performance targets, there must be a balance between being realistic and being ambitious.
- For the contracting utility, a well-designed PBC is less risky since it will pay the full price of the contract only if targeted performance improvements are achieved.

However, a PBC needs to be embedded in a long-term sector vision (at national and at utility level). Financing of OPEX and continuity of CAPEX (recurrent 3R – repair, renew, replace investments) are key for sustainable results.

PBCs in the water sector are still quite few, and rather recent, typically less than 15 years. The most frequent type of contracts, which are widely used, are for non-revenue water (NRW) reduction projects. Quite recently, performance-based contracts have appeared as a suitable type of contracting by municipalities willing to partner with a private company to reduce non-revenue water, a major challenge in several big cities around the globe.

Performance-based contracts have been favored for now about 25 years for procurement in many sectors, including in public procurement, especially in the USA. The US General Accounting Office issued a report in 2002 proposing guidance for performance-based service contracting. This report concluded that performance-based contracting offers a viable way toward achieving savings and getting better results from contractors.

In 2010, OECD issued guidelines for performance-based contracts between water utilities and municipalities. The report brings strong insight on best practices to design performance indicators and provide useful recommendations for indicator monitoring so that the contract can execute well. The guidelines ponder that "bonuses are best applied with management contracts, penalties are mostly suited for lease contracts while regulating risk through tariff adjustment can be the preferred option in both lease and concession contracts".

Most PBCs are hybrid schemes, combining fixed payment with variable payments. While a 5–10% bonus provides some incentives for performance, a "true" PBC shall aim for at least 20 or 30% of the remuneration paid through variable fees – so as to include not just profit but also part of the contractor's costs (see Figure 1.2). The feasibility of having at least 20% paid through bonus depends on local conditions and contractual design. Allocation of risks must be carefully thought about together with contractual enhancements to mitigate risks (e.g. donors support with partial risk guarantee (PRG)). PBCs require "upfront" financing since they link remuneration to results.

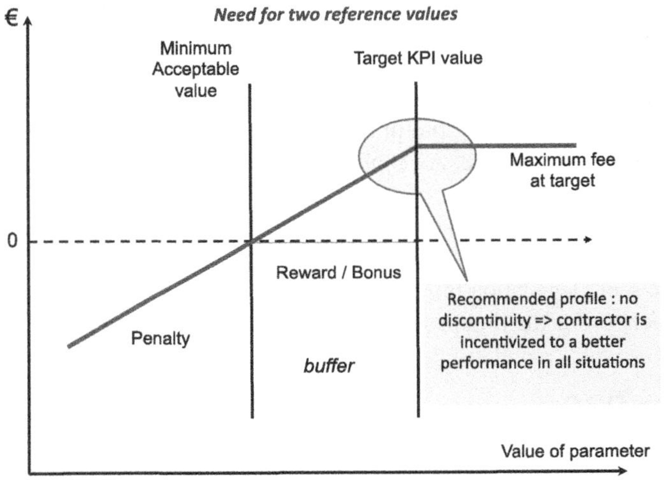

Figure 1.2 Bonus–penalty incentive scheme in a PBC.
Source: Carron *et al.* 2012.

Validated baseline information is thus an absolute precondition for an effective incentive structure with realistic time-bound performance targets. Inaccuracy of baseline is an issue and a constraint for a successful PBC. Such a baseline requires time and money. This leads to the concept of sequential, or progressive performance-based contracts, evolving from a contract of means (input based) (during which the baseline data are validated) towards a contract of results (output based). Such an intermediate contract (input based) first step may be needed during which also institutional, operational, and/or financial viability questions are addressed. But the risk may be that a stepwise approach may never go beyond the first step, and may involve complex re-bidding issues.

1.2 REFERENCES

Carron, D., Cavard, J. and Cohen, A. (2012). "Integrating performance targets in contracts between communities and contractors - The SEDIF case study. Improving Performance of Water and Sanitation Public Services, Astee Book, Edited for the 6th World Water Forum in Marseille (2012)."

EU Commission. (2013). European Energy Efficiency Platform E3P. See https://e3p.jrc.ec.europa.eu/articles/esco-financing-options (accessed May 2018).

Chapter 2

PBC and results-based financing: The inverse approach

Didier Carron

Reducing non-revenue water (NRW) typically is made through a panel of various actions best addressing the physical and the commercial losses estimated from an initial baseline. The balance between these actions, which may vary every year, is however not easy to attain to provide efficient and sustainable results at the lowest cost. The inverse approach is a tool to calculate such an optimal set of actions and can be used either through a financial-resources driven approach or an NRW reduction target driven approach according to financial and time existing constraints.

2.1 INTRODUCTION

PBCs have two major differences with standard contracts:

- Remuneration of the contractor is at least partially based on performance and not on time or through a lump sum.
- The contract, being based on an obligation on the outcome, might not define the means that the contractor will use to reach the objective.

As a result, at the heart of a PBC, is the definition and the setting of the objectives. Clearly:

- The objectives should be reachable
- The objectives should be financially calibrated so that:
  - The contract is nor too risky nor too profitable for the contractor
  - The possible partial funding by the public authority, potentially backed by an international financing institution (IFI), is adjusted.

There is thus the need in an early phase to have a first good guess at the cost of the project for the objectives in order to both calibrate the objectives and prepare the financing of the project. The cost of

Authors: Jan Janssens, Didier Carron, Philippe Marin, Philip Giantris and Tom Williams
doi: 10.2166/9781780405964_0007

the project is the sum of the investment costs and the operation and maintenance costs, and does not take future benefits of the project into account.

In developing countries, part of the investment can typically be funded by an IFI, however IFIs need rapidly to get a rough estimate of the amount of the investment to finance, which will determine the amount of the loan to be provided.

To achieve long-term benefits for the project, there is the need for investment which will bring a return in the future.

When applied to non-revenue water (NRW), actions to take in order to reduce the non-revenue water are numerous and do not yield the same results as a function of time. Immediate (i.e. less than 2 years for implementation) operating actions will bring fast results but they might be limited in time and/or insufficient to reach the targets. On the contrary, investments like renewal of pipes will be expensive and bring visible results only in the long-term.

In fact many factors influence non-revenue water. If the influence of each factor can be guessed using models and past experience, the combination of these factors over the long-term is typically ignored, leading to favoring of short-term solutions which are not the most efficient nor the most cost-effective over the long-term.

The list of main actions which can be implemented to reduce non-revenue water is as follows:

- Commercial losses (apparent losses)
  - Metering inaccuracies → renewal of meters
  - Illegal connections → detect and replace by legal connections
  - Unauthorized consumption → detect and replace by legal supply (water points, etc.)
  - Unbilled consumption or uncollected bill → reinforce customer management and debt collection
- Real losses
  - Pressure regulation
  - Network logging for global leak detection
  - Design of districted metered areas (DMAs) for more efficient leak detection
  - Pipe inspection
  - Monitoring of night flows
  - Active leak detection and leak repair
  - Renewal of pipes (house connections and network)
  - Renewal of valves

Among these actions, the renewal of pipes is very difficult to handle. One sees that some replacement is needed, as time can only deteriorate the network conditions, as demonstrated by any natural rate of rise in leakage (NRR) analysis.

UKWIR (2005) explains that "NRR is essential for determining the intensity of active leakage control that is required to maintain leakage at a specific level. As an indicator of asset condition, NRR is also central to mains replacement strategy". This clearly states the balance to make between short-term actions and renewal of pipes, as NRR increases with the years.

Although pipe replacement algorithms can propose a prioritized list of pipes to replace – typically based on an analysis of bursts and not of leaks – they do not provide quantitative indications on the expected effect of replacement.

Also some short-term actions can also have long-term effects, for example, pressure regulation. Pressure regulation reduces the leak flow as soon as implemented, but also slows down the degradation of pipes, contributing to increasing their life expectation and reducing the need for replacement.

This shows that:

- There is the need to know how much of each action should be taken every year to reach the targets in a cost-effective and sustainable manner.
- There are dependency links between actions, both in the short-term and in the long-term, and their efficiency depends of the asset conditions. For example:
  - Pressure regulation has long-term effects which can only be taken into account over a long period of time
  - Active leakage control is more efficient if DMAs have been created and monitoring of night flows is performed
  - Active leakage control is less efficient when physical losses are low, such as for example close to the UARL as defined by Lambert.

2.2 THE DIRECT PROBLEM

A model predicting the NRW over the next period of 10 or 15 years has been developed. It calculates the NRW water every year for a given action plan defined over a long period of years (typically 10–20 years). The NRW calculation for a year depends on actions already taken in previous years (DMAs created, pipes replaced, illegal connections suppressed, etc.), on the conditions of assets and on the NRW level at the beginning of the year (Figure 2.1).

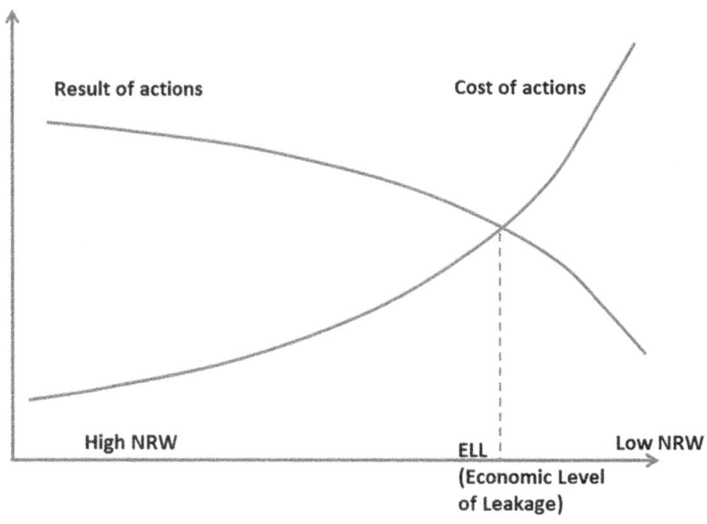

Figure 2.1 The relationship between, actions, costs and NRW.

Standard ratios are used in this model, which of course are ideally replaced by more accurate and locally relevant rules or equations derived from available data when it exists, which will better represent the effect of each action in the case under study.

However, the difficulty is not so much the uncertainty regarding these ratios or equations as the fact that they depend on the conditions of the network, i.e. on the NRW level, introducing additional complexity and uncertainty.

This means that "dynamic" relations (equations) must be used in the modeling.

Input data required is:

- Technical data:
  - Water balance at year 0, with volumes for each category and subcategory
  - List of pipes and connections, with characteristics. A prioritized list of pipes to replace is helpful to have
  - List of meters with characteristics. Any data on inaccuracy from meters is appreciated in order to get an estimation of meter conditions and choose the right ratio to estimate the inaccuracy
  - List of valves with characteristics. Any data on valve conditions (leaks in the last years, inspection, etc.) is appreciated in order to choose the right ratio to estimate their risk of failure
  - NRR estimate, to be calculated separately from existing NRW data, otherwise will be proposed using ratios
  - Average pressure in the network, or pressure for each existing sector if any.
- Economic data:
  - Cost of every action in the local context (investment and operation and maintenance).
- Efficiency of actions:
  - Volume of water saved per action. This volume depends itself on the actual NRW components. It can be estimated from analysis of previous actions, or, in case of no data, ratios observed in similar conditions can be used.

Based on this data, the model calculates, for a given action plan, made of a set of actions every year over a period of years:

- The NRW every year, from year 0 to final year N
- The cost of the action plan: investment and operation. Alternatively, a net present value (NPV) can be calculated.

Figure 2.2 illustrates the global logic of this model, called "the direct model".

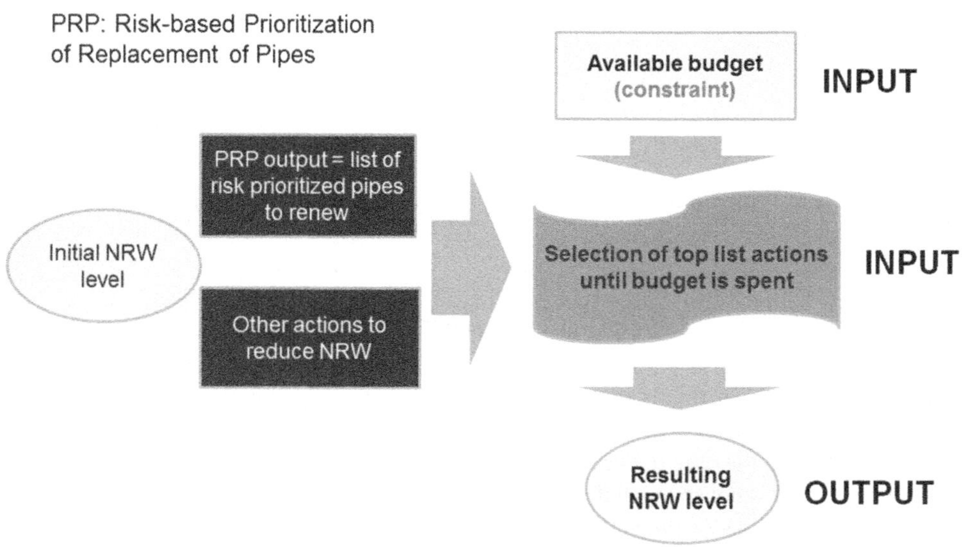

Figure 2.2 The global logic of the direct model.

2.3 THE INVERSE PROBLEM

The inverse problem (Carron, 2016) reverses the roles of the input data and the outcome of the model:

- The outcome is now the input data: the targeted NRW year per year is set
- The action plan becomes the output

There are an infinity of action plans able to provide the desired result. What is searched for is the action plan with **minimum cost**.

Thus the inverse problem is stated as follows: "Find the action plan with the minimum cost able to provide the targeted NRW according to the direct model."

The inverse model can thus also be called a *result-based approach*.

From a mathematical point of view, the targeted level of NRW every year becomes a constraint in the algorithm which minimizes a cost function (mixing operation cost and investment cost) over a long period (10–20 years), making it possible to find the most cost-effective set of NRW reduction actions, minimizing OPEX and/or CAPEX.

The inverse approach can be led under two schemes:

The financial resources driven approach

A given amount of money (financial resources) is available, and one has to find the best action plan to have the maximum reduction of NRW

or

The NRW reduction target driven approach

A given NRW reduction is targeted (for instance, from a financial model providing expectations in revenue from the reduction of NRW);

The scheme for the available financial resources driven option can be represented as in Figure 2.3.

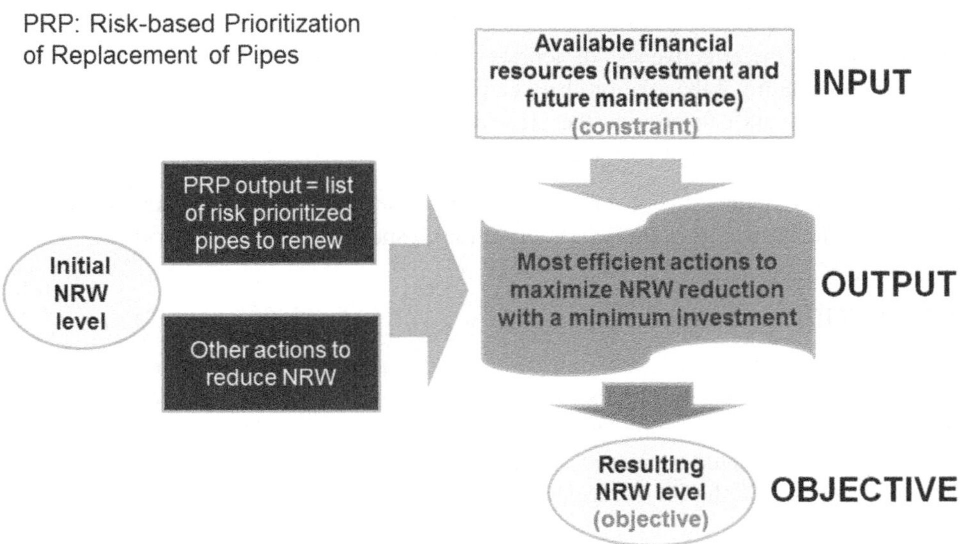

Figure 2.3 The available financial resources driven approach.

The scheme for the NRW reduction driven option can be represented as in Figure 2.4.

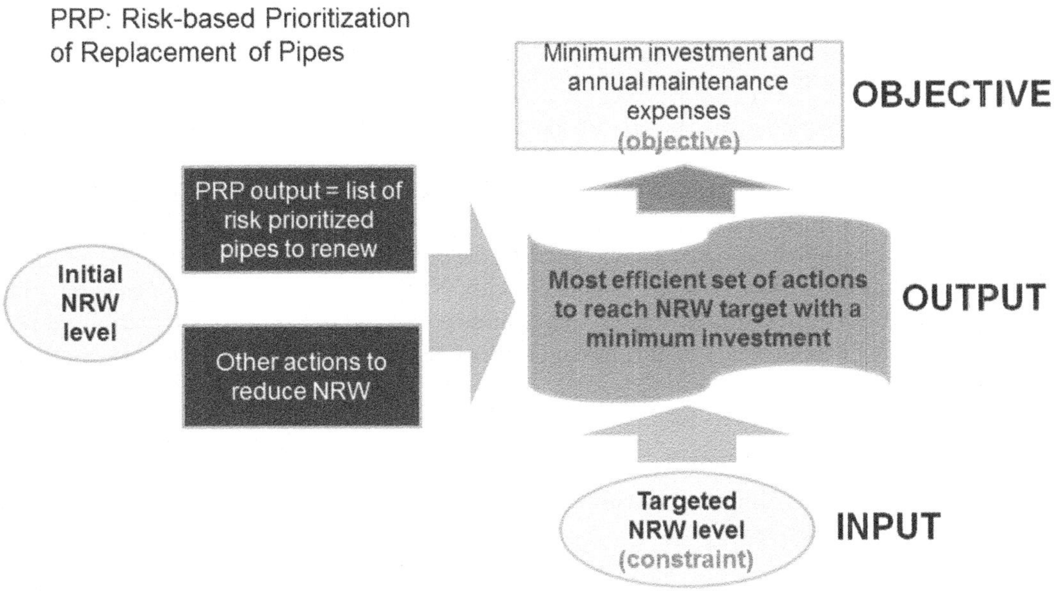

Figure 2.4 The NRW reduction target driven approach.

Calculating the optimal NRW action plan faces two difficulties:

(1) An action plan is made of 10 to 20 variables every year (length of network logged, number of illegal connections replaced, etc.), so this represents several hundreds of variables over 20 years, which is quite high.
(2) In addition, as the effectiveness of actions depends on actions taken the previous years, due to the dynamic nature of the relations between costs and results of actions, only a global calculation including all years can provide the effective performance of an action plan. This makes the inverse problem non-linear in mathematical terms.

As a consequence, solving this inverse problem and finding the optimal solution requires quite sophisticated algorithms, such as evolutionary algorithms, especially "genetic algorithms".

2.4 CASE STUDY

An example is shown in Figure 2.5 for a utility comprising about 11,000 connections, a network with a total length of pipes and connections of about 400 km, with an average age of pipes of 55 years.

The initial NRW level is 23% and the targeted level is 17%, corresponding both to a regulatory requirement and to the estimated economic level of leakage (ELL), although calculated very roughly.

The targeted level of 17% should be reached within 6 years, i.e. in 2021, while the initial state is 23% in 2015.

Results when running the inverse model for the first 10 years of a period of 15 years are shown in Figure 2.5.

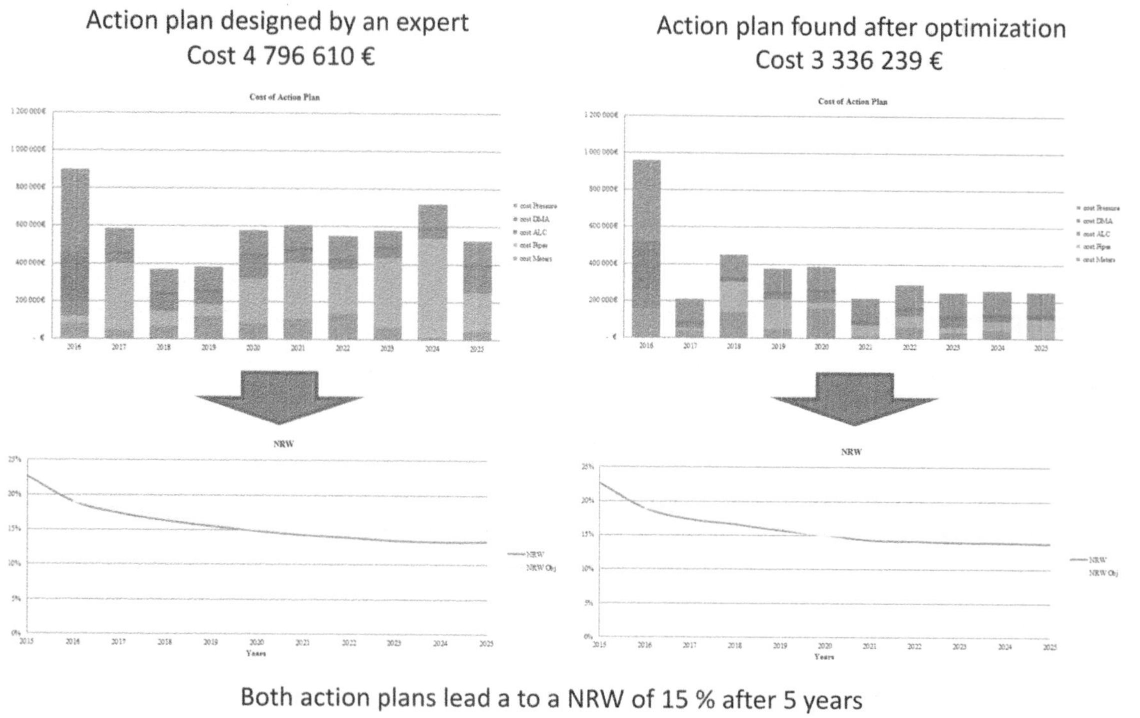

Figure 2.5 Results for the case study when running the inverse model for the first 10 years of a period of 15 years.

Interestingly, the method proposes a best solution which includes several types of actions every year, including pipe replacement.

It thus truly finds the best compromise between short-term actions and more costly but long-term sustainable actions such as pipe replacement, due to the fact that the global cost over 10 years is taken into account.

The total cost of the action plan over 5 years is below €5M, whereas a first guess of actions was about €10M. It thus shows that by running the inverse problem, a much more cost-effective solution could be found, saving half of the initially expected cost while being sustainable.

2.5 CONCLUSION

The result based approach makes it possible to design action plans to ensure reaching a targeted NRW level in a sustainable manner.

- It can be used by institutions to calibrate their support.
- It can be used by Water Utilities to evaluate feasibility of NRW targets and its associated cost.
- It can be used by Operators and Contractors to design the most effective action plan to reach an NRW level target.

It is most likely useful to run such an approach when planning to use a performance-based contract, in order to calibrate the expected cost of the contract according to objectives.

It can also be used again later, when a PBC tender is ongoing and a proposal needs to be evaluated, in order to challenge the mix of actions proposed by the bidders.

For sure, the outcome of the model (i.e. the action plan), when based on standard ratios to calculate the effect of actions, is approximate. However, the model can be used in a dynamic way, by adjusting these ratios or substituting them by more suitable equations, when more data becomes available with the years.

In case of additional constraints, such as for example a cap on annual investment or time constraints for implementing actions, they can easily be entered into the model "by hand", so that the proposed optimum action plan complies with these constraints.

2.6 REFERENCES

Carron D. (2016). Long term non-revenue water and network management. *Poster presented at the 2016 IWA World Water Congress & Exhibition*, Brisbane, Australia.

UKWIR (2005). *Natural Rate of Rise in Leakage*. Ref: 05/WM/08/33, UKWIR, London.

Chapter 3

PBC and energy efficiency

Didier Carron

Performance-based contracts (PBCs) for energy saving are common in the energy and in the building sectors. Energy savings performance contracts (ESPCs) or more simply energy performance contracts (EPC) are often used in order to accelerate investment in cost effective energy conservation measures without the need for up-front capital from the requesters. Such contracts are typically signed between the requester and an energy service company (ESCO). Energy is however also emerging as a major issue for water utilities. They can gain experience from these contracts to address their issues on energy saving and energy efficiency and implement energy conservation measures (ECMs) either through shared savings performance schemes or through guaranteed savings performance schemes.

Several case studies demonstrate how effective such approaches can be for water utilities.

3.1 INTRODUCTION

As early as 1992, in the United States, the Energy Policy Act authorized Federal agencies to use private sector financing to implement energy conservation methods and energy efficiency technologies.

Since then, most countries have implemented energy savings performance contracts (ESPCs) or more simply energy performance contracts (EPCs), in order to accelerate investment in cost effective energy conservation measures without the need of up-front capital by Federal agencies.

Such contracts are typically signed between the requester and an energy service company (ESCO). The ESCO conducts a comprehensive energy audit for the Federal facility and identifies improvements to save energy. In consultation with the Federal agency, the ESCO designs and constructs a project that meets the agency's needs and arranges the necessary financing. The ESCO guarantees that the improvements will generate energy cost savings sufficient to pay for the project over the term of the contract. After the contract ends, all additional cost savings accrue to the requester.

Authors: Jan Janssens, Didier Carron, Philippe Marin, Philip Giantris and Tom Williams
doi: 10.2166/9781780405964_0015

However these contracts have been mainly used so far in the building sector, although there are typically no restrictions for their use in other sectors.

In the water sector, James Horne, from US EPA Office of Wastewater Management, reported that water and wastewater treatment represents about 2% of the nation's energy consumption, and that energy costs are often one of the top 3 O&M costs in this sector (Horne *et al.* 2016). According to James Horne, several studies estimate potential savings of 15–30% that are "readily achievable" in water and wastewater plants, within payback periods of only a few months to a few years. This shows that there is room for ESPCs also in the water sector.

The Office of Water of the US Environmental Protection Agency (EPA) has issued a guideline in July 2013 on strategies for saving energy at public water systems (USEPA, 2013). It is reported that at the national level, drinking water and wastewater systems account for 3–4% of US energy use. This is equivalent to 56 billion kilowatt hours (kWh) annually, and the generation of almost 45 million tons of greenhouse gases (GHG). At the community level, drinking water and wastewater systems are typically the largest energy consumers accounting for 25–40% of a municipality's total energy bill. Approximately 80% of municipal water processing and distribution costs are for electricity.

In the European Union, Directive 2006/32/EC on energy end-use efficiency and energy services has defined energy performance contracts as contractual arrangements between the beneficiary and the provider (normally an ESCO) of an energy efficiency improvement measure, where investments in that measure are paid for in relation to a contractually agreed level of energy efficiency improvement. ESCOs are defined as natural or legal persons that deliver energy services and/or other energy efficiency improvement measures in a user's facility or premises, who accept some degree of financial risk in so doing. The payment for the services delivered is based (either wholly or in part) on the achievement of energy efficiency improvements and on the meeting of the other agreed performance criteria. These tools are thus quite similar to energy savings performance contracts.

3.2 ENERGY SAVINGS FROM PUMPING

Pumping is typically by far the largest energy consuming function within a supply water or a sewage system. Energy savings can be made through several actions such as:

- Proper design for new equipment: pick pumps to operate efficiently at conditions where the pump will actually operate most of the time
- Consider variable speed drives or smaller pumps to improve efficiency at low flows
- Operate "smart pumping", including pumping as slowly as possible
- Utilize storage to level out pumping rate
- Replace old motors
- Replace pumps: it has to be remembered that the initial cost of a pump is typically about only 10% of its life cost, whereas energy costs and maintenance costs associated with that pump are 45% and 37%, respectively (ESMAP, 2011).

A systemic approach can be used to tackle this issue:

- Pumping scheduling: use pumps at hour when energy tariffs are lower (typically during the night)
- Use of an energy management system, able to forecast demand and adjust to the most cost effective pumping profile

- Revisiting of a global pumping scheme: in intermeshed networks with many interconnected storage tanks fed from various plants, pumping schemes can be optimized using fancy algorithms such as genetic algorithms.

Energy savings for pumping up to 20% can be achieved by implementing these actions, with a payback period typically not more than a few years.

3.3 OTHER ENERGY EFFICIENCY (EE) POSSIBILITIES

Other energy savings can be obtained through other actions such as for example:

- Production of renewable energy from solar panels or microturbines
- Optimization of travels by fieldworkers
- Decrease of reactive power
- Heat recovery from water /wastewater
- Biogas production from treatment sludge
- Energy consumption shading

Many technical guides and handbooks are nowadays available on this subject.

3.4 PERFORMANCE-BASED CONTRACTS FOR ENERGY EFFICIENCY

Like the reduction of non-revenue water (NRW), energy efficiency (EE) is very well adapted for performance-based contracts (PBCs) as an energy efficiency project will save costs and potentially generate additional revenue.

In a technical report on Energy Efficiency for Municipal Water and Wastewater Utilities by the Energy Sector Management Assistance Program (ESMAP) administered by the World Bank, issued in February 2012 (Liu *et al.* 2012), it is reported that:

> *IFC is considering investment in several specialized companies that provide efficiency services to water utilities. These companies identify energy savings opportunities through NRW reduction, water pump efficiency optimization, WWTP aeration optimization, and other options. They typically operate on a performance-based contracting model in which a portion of capital expenditure contribution for improvements is financed by the ESCO and a portion of the revenue is linked to meeting energy and/or cost reduction targets. [...] Financing hurdles can be reduced by introducing dedicated EE funds to address large but disaggregated investment needs and by promoting third-party financing through energy/water savings performance contracts.*

The concept of "energy savings performance contracts" (ESPCs) is introduced: an ESPC is defined as "an energy service company (ESCO) that provides an energy consumer or 'host facility' a range of services related to the adoption of EE products, technologies, and equipment. By procuring external EE services, a WWU gains much needed technical expertise". It is mentioned that "The more innovative part of ESPCs is that they can double as financing instruments, in addition to being energy management tools. Full ESCO services may include financing for the EE upgrades, disencumbering the host facility from the burden of securing upfront capital".

As is well described in this report as well as in the "Energy and Water Efficiency in Municipal Water" report by The Alliance to Save Energy/Watergy (Barry, 2007), performance-based energy efficiency contracts (PBEECs) can be of three types in the water and wastewater sector:

- Shared Savings Performance Contract: a full service ESCO designs, implements (equipment and software), verifies and gets paid from actual energy saved.
  - The savings are shared between the utility and ESCO with the contract stipulating that the utility will receive a certain percentage of the savings, but it does not guarantee the magnitude of those savings.
  - All risk lies with the ESCO. The ESCO takes on the risk of any third-party financing from a lender, putting the loan on the ESCO's balance sheet.
- Guaranteed Savings Performance Contract: an ESCO with third-party financing: designs and implements projects, and guarantees a minimum level of savings
  - The ESCO guarantees that savings will be sufficient to cover the investment cost, and if they are not the ESCO pays the difference between the realized savings and project payments.
  - The guarantee can be either a one-time payment or on-going.
  - If external financing is needed, the utility takes on the third-party financing from a lender, putting the loan on the utility's balance sheet.
  - Excess savings can be shared between the utility and the ESCO.
- First Out Performance Contract: an ESCO acts as a full-service ESCO, but the contract term varies based on actual savings (variable-term contract).

Figure 3.1 below illustrates a typical impact for a water utility when contracting with an ESCO company:

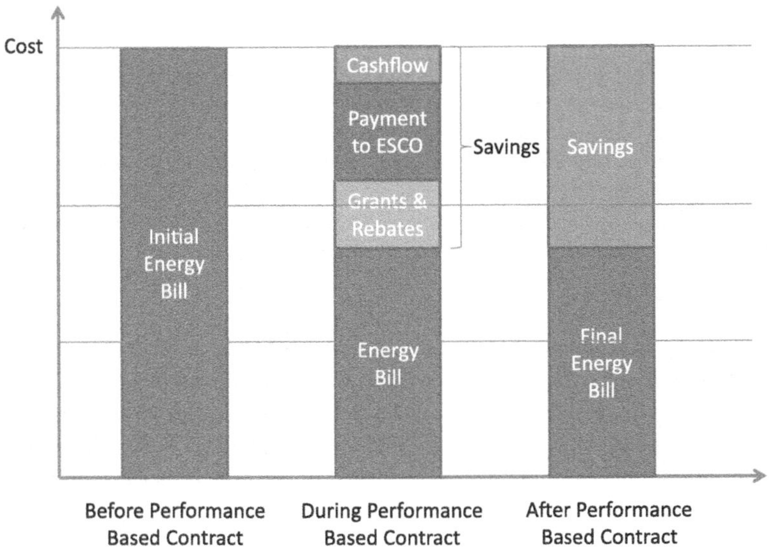

Figure 3.1 Financial impact for a water utility when engaging a PBC with an ESCO.

3.5 DESIGNING AND IMPLEMENTING AN EE PERFORMANCE-BASED CONTRACT

The implementation of an EE performance contract follows the same general scheme as other PBCs.

Figure 3.2 by the USEPA (2013) illustrates the general scheme to use.

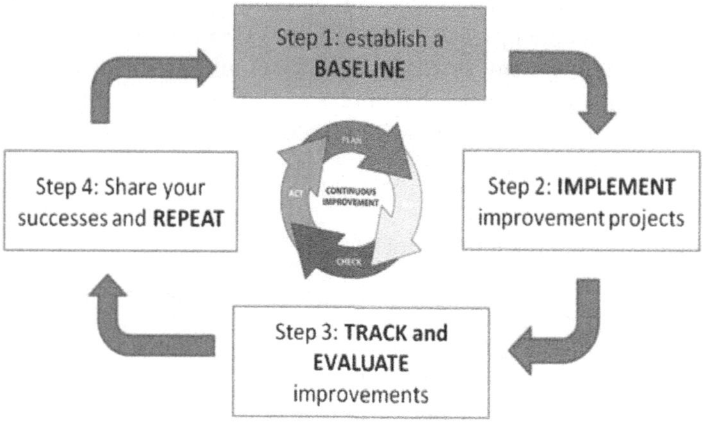

Figure 3.2 Steps for an energy improvement program – strategies for saving energy at Public Water Systems – US EPA.

Step 2 on the above figure is the implementation of energy conservation measures (ECMs). Once a drinking water system has determined its baseline energy use and conducted an energy audit, the next step is to identify, evaluate and prioritize potential ECMs. Typical criteria used to prioritize ECMs include:

- Estimated capital or upfront investment
- Estimated O&M cost reduction or increase
- Expected energy reductions
- Calculation of simple payback periods – the number of years of energy savings that it will take to account for the costs of the energy efficiency improvement
- Annual cost savings (both energy and O&M).

KPIs used to establish the performance based mechanism within the performance-based contract can include the following:

- Energy efficiency improvement: can be calculated globally or per metre cubed processed
- Global energy cost reduction. Can be calculated globally or per metre cubed processed
- Energy produced in kilowatt hours (kWh)
- Return on investment

3.6 CASE STUDIES

3.6.1 Yerevan (Armenia)

Source: ESMAP (2011)

In 2000, the Armenian capital's water utility, the Yerevan Water and Sewerage Enterprise (YWSE), entered into a 5-year, performance-based management contract with private operator Acea Spa Utility (Acea).

The contract included three main targets:

- Improvement of metering and of collection rate
- Increase of water supply from 2 to 6 h per day to at least 10 h per day
- Reduction of electricity consumption by at least 20%

The entire project cost US$28.91 million, including capital investments, management contract fees, and operator bonuses. During the contract period, Acea disbursed US$24.07 million of the capital investment fund to rehabilitate the existing system. It was paid US$3.43 million for managing YWSE for 5 years and received a US$1.41 million performance-based bonus. The capital investment fund and management contract was financed in part from a US$30.87 million credit from the World Bank.

The results achieved are shown in Table 3.1.

Table 3.1 Key performance indicators for the project.

Key Indicators	Baseline (2000)	Target (2004)	Achieved (2004)
Increased daily hours of water supply	6 h/d	10 h/d	18 h/d
Reduced power consumption	240 GWh	192 GWh	169 GWh
Increased gravity use	40%	NA	52%
Decreased NRW	73%	48%	80%
Increased metering	1,000	20,000	277,000

In particular, the "energy efficiency investments" were deemed highly cost effective. The simple payback period of the energy efficiency investments was estimated at 3.5 years, as shown in Table 3.2. This was evaluated using the annual electricity cost savings (US$4.83 million/year) and all investments that had energy-saving impact (US$16.78 million) based on data from project documents. This represents a conservative estimate, as some of the costs had benefits beyond energy efficiency and many of the benefits accrued went beyond mere energy savings.

Table 3.2 Payback of investments costs.

Cost-Effectiveness of Energy Efficiency Measures	
Benefits: average annual electricity savings	US$4.83 million/year
Energy efficiency investment costs	US$16.78 million
• Sectorization costs (61% of the capital investments) = US$14.68 million	
• Pump upgrades and leaky pipes rehabilitation + gravity use = US$2.10 million	
Simple payback	3.5 years

3.6.2 Emfuleni (South Africa)

Source: Barr (2007), Alliance to Save Energy (2006), C40 Cities Climate Leadership Group; McKenzie and Wegelin (2005)

The areas of Sebokeng and Evaton are part of Emfuleni Municipality in South Africa, about 50 km south of Johannesburg. The population is about 450,000 residents, and the number of household connections about 70,000. Before the project, non-revenue water was estimated at about 75%–80%, with very high energy consumption per metre cubed delivered to customers.

The municipality decided to launch a performance-based contract in order to regulate the pressure during off-peak periods, in order to reduce water lost through leakage and decrease energy consumption. The contract was a mix of performance contracting based on shared savings and a build–own–operate–transfer (BOOT) type arrangement. This contracting mechanism was deemed to be the best approach in light of the municipality's limited ability to access capital and its lack of technical capacity with regard to pressure management.

The municipality decided that the contract would be 100% performance based, the payments to the Contractor being based on 20% of the savings achieved with an additional safeguard of a monthly "cap".

The funding for the investment (US$0.5 M) was fully provided by the Contractor, with no upfront cash flow needed by the Municipality. However, the Municipality had to be assisted by consultants and pressure regulation experts, at its own cost.

The following was expected from the contractor:

- Financing and installation of a pressure regulation system
- Operation and maintenance of the system
- Utility staff training so that it becomes able to manage the system at the end of the contract

The contract was established for a duration of 5 years, so that the payment to the contractor can allow him to recover its investment and make a profit. After this period, 100% of the savings were kept by the Municipality.

The results were outstanding: the savings during the 5 years of the contract amounted to more than 50 million m^3 in water purchases and annual energy savings around 14,250 MWh.

The payback period appeared to be less than 2 years for the contractor and less than 4 months for the Municipality.

3.6.3 Larger contracts including an EE performance based component

Many performance-based contracts between utilities and contractors can include an energy efficiency component. This is even frequently the case with management contracts, which define a set of key performance indicators (KPIs) with targets and include energy efficiency as one of these KPIs.

Below are two examples of incorporating energy management into management contracts.

Source: EPA.

3.6.4 Milwaukee

The existing owner/operator contract between Milwaukee Metropolitan Sewer District (MMSD) and Veolia Water Milwaukee (VWM) stipulates that the operator (VWM) contributes 25% to the system's energy costs. This arrangement incentivizes both parties to actively seek ways to reduce energy costs. The contract also states that cost savings garnered from energy-related projects initiated by VWM and implemented by MMSD will be shared 50–50. While sharing savings can create further incentives to work together on energy projects, it has made financial modeling more difficult. The two parties recognize that further clarification on savings sharing is needed in future contracts.

3.6.5 Lynn

In 2001, the Lynn Water and Sewer Commission entered into a 20-year operations contract with Veolia Water. The contract contains explicit language on how energy-related operating costs, capital costs, and net savings will be shared. It stipulates that Lynn Water and Sewer will be financially responsible for a maximum of 13,159,000 kWh per year, where any kilowatt hours in excess of the cap will be the responsibility of Veolia. If annual usage is less than 13,159,000 kWh, both parties equally share in the savings. The cost of energy improvements are shared 50–50, as well as the net savings incurred from the improvements.

3.7 REFERENCES

Barry J. A. (2007). Alliance to Save Energy, Water Case Study, Emfuleni Municipality, South Africa, May 2006.

Barry J. A. (2007). Watergy: Energy and Water Efficiency in Municipal Water – Supply and Wastewater Treatment: Cost-Effective Savings of Water and Energy. The Alliance to Save Energy, February.

C40 Cities Climate Leadership Group. Best Practices – Water: Emfuleni, South Africa, Clinton Climate Initiative. See http://www.c40cities.org/bestpractices/water/emfuleni_efficiency.jsp (last accessed July 2017).

EPA Designing RFPs and Contracts that Promote Energy Improvements at Contract-Operated Drinking Water and Wastewater Utilities Energy and Global Climate Change in New England, EPA Region 1 (New England), Presentation available on web site https://www3.epa.gov/region1/eco/energy/designing-rfps-contracts.html (last accessed February 2018).

ESMAP (2011). Energy Sector Management Assistance Program, Good practices in City Energy Efficiency, February 2011.

Horne J., Turgeon J. and Byous E. (2016). Energy Self-Assessment Tools and Energy Audits for Water and Wastewater Utilities, US EPA Office of Wastewater Management. Presentation available on https://www.epa.gov/sites/production/files/2016-01/documents/nrwa-energy-audits-for-small-utilities-8-4-14.pdf (last accessed July 2017).

Liu F., Ouedraogo A., Manghee S. and Danilenko A. (2012). A Primer on Energy Efficiency for Municipal Water and Wastewater Utilities. Energy Sector Management Assistance Program (ESMAP), February.

McKenzie R. and Wegelin W. (2005). Sebokeng/Evaton Pressure/Leakage Reduction: Public–Private Partnership. Conference Proceedings, Leakage. See http://www.wrp.co.za/about/articles/ppp-deliver-advanced-pressure-management-success-south-africa (last accessed July 2017).

USEPA (2013). Strategies for Saving Energy at Public Water Systems. Office of Water, US Environmental Protection Agency, July 2013.

Chapter 4

Internal performance contracts: A case of the National Water and Sewerage Corporation in Uganda

Marine Colon and Silver Mugisha

This chapter aims at sharing the lessons learnt from the implementation of performance contracts in the Ugandan water sector. These contracts have been signed since 2000 between the public water and sanitation operator, the National Water and Sewerage Corporation, and its supervising ministries. In addition, from 2004, internal performance contracts have been introduced between the NWSC and its operating agencies. The authors present the first main achievements and main characteristics of these contracts. They then outline the key lessons to prevent ill effects in the future. This analysis draws the link between performance contract design and the management style and culture of the public enterprise. It is based on two complementary insights: the experience of Silver Mugisha, a water engineer and Doctor of Economics, who took part in the reform process and is now the MD of NWSC; and a three-year research project led by Marine Colon, a water engineer and researcher in management sciences.

4.1 INTRODUCTION: WHY THIS CHAPTER?

The National Water and Sewerage Corporation (NWSC) is an interesting case to learn about when it comes to new public management, and especially public–public partnerships. A performance contract (PC) was introduced in 2000 between NWSC (the state-owned utility) and its supervising ministries (Ministry of Finance and Ministry of Water and Environment). The same year, the managing director introduced internal performance contracts (IPCs) between the headquarters of the utility and its branches in charge of operations, called "areas". The IPCs have become the spine of the management at NWSC, and seem to have greatly helped the utility to become a strong organisation compared to its original status when it used to be small, heavily indebted and less efficient.

The NWSC has indeed an outstanding history in the water sector. The state-owned enterprise was created in 1972 to operate the water and sanitation infrastructures in the three biggest cities (Kampala, Entebbe and Jinja). Its mandate enjoins it to be financially sustainable, i.e. water sales are supposed to fund all costs (operations, depreciation and capital). However, two decades of political unrest (dictatorship

 Performance-Based Contracts (PBC) for Improving Utilities Efficiency: Experiences and Perspectives
Authors: Jan Janssens, Didier Carron, Philippe Marin, Philip Giantris and Tom Williams
doi: 10.2166/9781780405964_0023

and civil wars) in the 1970s weakened its infrastructure and institutional capacity to deliver services. The rebuilding started at the end of the 1980s, including staff training and infrastructure development. From the middle of the 1990s, a focus on performance, financial sustainability and personnel incentives were the first step towards corporatisation. From 2000, the NWSC adopted a result-based management approach with the adoption of PCs and IPCs. Currently (2016), NWSC operates most of the large urban areas of the country (162 against 3 in 1972 and 9 in 1995) (see Table 4.1 below), supplying nearly 7 million people. It funds all of its operating costs (including depreciation) and a portion of capital investments. For this remarkable path, in the East African context, the management of NWSC has been awarded several distinctions, such as the IWA prize for "Outstanding Contribution to Water Management and Science" in 2010, Africa Utility of the Year (2014, 15 and 16), Best Government Procurement Complying Entity (2015), Best Digital Customer Care Entity (2015), among others.

Table 4.1 Evolution of the performance of NWSC from 1995 to 2016.

Indicator	Description	1995	2000	2005	2010	2016
Number of towns operated		9	12	15	23	162
Population to be supplied ('000)		885	1035	2290	3000	7000
Number of water connections		24,196	66,234	123,046	246,259	465,000
Number of sewerage connections		12,503	13,010	13,624	15,561	19,705
Water coverage (%)	Percentage of urban population with water supply	39%	57%	68%	74%	78%
Non-Revenue Water (%)	Percentage of water produced that does not generate revenue	68%	43%	34%	33%	29%
Collection efficiency (%)	Percentage of amount billed that is collected	63%	75%	89%	98%	102%*
Turnover	(billion UShs)	20	26	54	112	265
Labour productivity	Number of permanent staff per 1000 water connections	43	16	9	6	6

Source: NWSC annual reports.
*Due to payment of the Government bodies' debt.

Clearly, PCs and IPCs were key in the NWSC success story. This case appears to be puzzling since performance contracts between two public entities have sometimes received inconclusive discourse from utility practitioners and policy makers. In the 90s, a wide study from economists for the World Bank concluded that performance contracts in the public sector were bound to be inefficient, mainly due to the inability of ministries in charge to exert control (Shirley, 1998). This chapter aims at shedding some light on this ongoing debate: what IPCs have brought, what were the pitfalls and achievements? This chapter presents a background of what are IPCs in the Ugandan context before moving to the key lessons.

4.2 IPCs AT NWSC: IDENTITY CARD

We should distinguish between two types of public–public partnerships in the Ugandan water sector (see Figure 4.1). Performance contracts (PCs) are signed between the head of the utility and the Board on one side and the ministries supervising the NWSC on the other side, i.e. the Ministry of Finance and the Ministry of Water and Environment. They give broad direction to the top management for 3 years, with an incentive mechanism that targets Top Management. Every year, the Top Management team receives a bonus computed by the Board, based on an agreed incentive formula, according to the actual performance achieved by the utility. Managers can earn up to 15% of their annual gross salary in incentives. Internal performance contracts (IPCs) are signed between the head of the utility and the head of its operating branches, which are called "areas". These contracts include three main elements: (1) a definition of roles and duties of both parties, (2) a level of performance to be achieved during a defined period of time, measured by a number of performance indicators, (3) an incentive mechanism. IPCs are designed as tactical plans for the utility to meet its overarching targets set in the PC. The IPC contracts are renegotiated annually and operationalised through action plans. The IPCs incorporate 20 performance indicators structured around selected key strategic priority areas (SPAs) cascaded from the PC. Of these indicators, five (5) are key to compute the management fee: non-revenue water (NRW), cash operating margin, working ratio, arrears and collection efficiency.

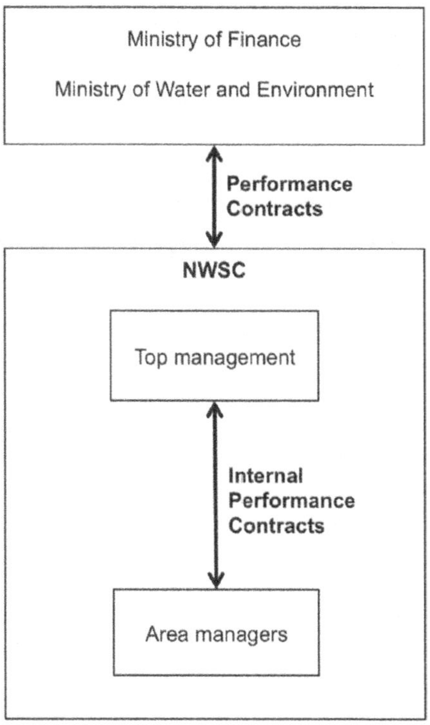

Figure 4.1 Contractual relationships at NWSC.
Source: Colon (2014).

One cannot understand how IPCs work without understanding the overall management and organisation of NWSC. Firstly, IPCs have contributed to and are now based on decentralisation of operations to the Area level. Each "area" is in charge of operating water and sanitation services in a dedicated urban centre like an autonomous body, under the supervision and support from the head office. The head office is in charge of policy-making, infrastructure development, and relationship management with donors and the Government.

Secondly, each "area" is headed by an "Area Manager" and his/her team of managers. Area managers are recruited through a bidding process aimed at selecting the best capacities to tackle the challenges of the area. The bidding process is based on preparation of an expression of interest, specifying a problem analysis, key strategies, targets to be met and resources required for implementation. From 2009, the bidding process has been open to any professional in Uganda, so even managers coming from the private sector are allowed to bid.

Thirdly, the incentive mechanism is combined with an equalisation to ensure solidarity between revenue generating areas like Kampala (70% of the overall revenue in 2010) and poorer areas. Areas are in charge of revenue collection, and send the revenue collected to an area-based bank account that is managed by the head office. The overall revenue generated is then redistributed to areas, to their specific operations bank accounts, under the form of a fee according to rules defined in the contracts. This management fee is used by areas to pay all costs except big capital investments. The fee pays staff expenses, operations and maintenance costs, administrative expenses, delegated capital expenditure and performance bonuses. Every month, performance achieved is compared to performance targets negotiated every year between the head office and the Area Managers and incentives determined accordingly. The incentive/bonus formula incorporates a mechanism for prorating between minimum base performance standards and targets. Weighted performance below minimum standards attracts penalties. Incentives can, potentially, go up to 40–50% of monthly staff salaries.

4.3 PREREQUISITES FOR IPC SUCCESSFUL IMPLEMENTATION: STRONG FOUNDATIONS AND COMPETITION WITH THE PRIVATE SECTOR

First of all, when IPCs were introduced (in 2000), the National Water and Sewerage Corporation was not starting from scratch. The utility was enjoying the benefit of nearly 15 years of refurbishment programs led under the then Managing Director (MD) Eng. Hillary Onek, a water engineer, by profession. Infrastructure was working, main operations standard operating procedures (SOPs) had been set up and the staff, especially engineers, had been well trained. The water tariffs had increased, initially to cover operation and maintenance costs. In 2004, a tariff indexation formula was introduced to cater for exogenous economic factors (inflation, foreign exchange and power costs) in order to keep the real value of the tariff. Consequently, with efficiency gains, the tariff is able to cover all operational costs including depreciation and some capital investments. Importantly, the legal framework had been set: the NWSC Statute promulgated in 1995 recognised the autonomy of management and operations to the utility. Therefore, the utility has become an independent public enterprise. The role of the government is mainly limited to policy formulation, appointing the members of the Board, vetting and approving the water tariffs proposed by the utility and negotiating the performance contracts. The NWSC Act requires the utility to be operated on a commercial basis aiming at financial sustainability. It seems unlikely that NWSC would have been able to build a result-oriented management without these foundations.

Secondly, the reasons for the introduction of IPCs seem to matter in their success. IPCs were introduced to put NWSC on a strong performance foundation for possible private sector participation (PSP). The Government had launched a large divestiture program funded by the World Bank planning to sell nearly all public companies to the private sector. At the end of the 1990s, a report by the World Bank pointed at the fragile financial situation of the utility, making it not yet ready for privatisation. In order to revitalise the performance of the utility, a new MD was appointed towards the end of 1998, Dr William Muhairwe, and a performance contract was signed in 2000 between the ministries of Finance and Water and

Environment with the top management of NWSC to urge the utility to improve its financial performance and expand the services. At that time, a management contract had been signed with a private company to operate the Kampala Area (1997–2001). Proving that the managers of NWSC were able to achieve superior performance to managers from western private companies has been a great motivational driver for internal reforms at NWSC. While looking like a private company may be controversial to some public organisations in Europe, in this particular case it has been quickly seen as a great improvement by the Government and the utility managers. Indeed, the public sector had been destroyed by political unrest, and was offering no good employment conditions or service delivery to Ugandan people. Before the reform, the NWSC was not able to pay salaries or to deliver quality and reliable services. Internal Performance Contracts have been useful to demonstrate that NWSC was also able to develop performance improvement programmes, implement and monitor them, thereby improving service delivery. After 4 years, the achievements of NWSC convinced the Ugandan Government that it was probably not necessary to privatise the utility – it was a question of good leadership and good change management practices. In this case, the divestiture project worked as a great driver to improve the performance of the public body. Successful IPCs provided the undisputable proof.

4.4 IPCs: AN INTEGRAL PART OF AN ORGANISATIONAL STRUCTURE, A MANAGEMENT STYLE AND AN ORGANISATIONAL CULTURE

Public-public partnerships are bound to fail when they do not fit the organisational culture or the way public organisations work. IPCs have been and are still useful to NWSC up to the present day for a number of reasons. The way IPCs were introduced and managed eased their adoption and contributed to profoundly changing the organisational culture. Before 1998, the utility was managed by engineers who were mainly focused on engineering issues, with little interest in commercial and customer management issues. The management style was pyramidal and centred around the head office, with little delegation of power and responsibility to the Area level. The corporation gradually introduced the following four key elements of IPCs, which contributed to the re-engineering of organisational culture at NWSC.

4.4.1 Key 1: Making managers accountable

The first key is to make staff and managers accept the idea of being accountable. NWSC used to espouse many characteristics of typical public water utilities, namely: centralised management, irregular staff attendance and late coming due to low salaries and poor customer and commercial orientation. The challenge was to switch to results-based management moving towards greater efficiency. Once again, the foundations were there. This started in the early 1990s with increased salaries (×10) against individual incentive schemes based on two very simple indicators: attendance and whether the job given is done. The next step at the end of 1990s was to get managers used to working with targets, taking additional responsibilities and striving to earn bonuses. Several programs were introduced for this purpose (Muhairwe, 2009). It took 5 years to achieve decentralisation of operations along with giving more responsibilities to Area Managers (1999–2004). Along the way, IPCs contributed to refining the roles of each part of the organisation.

Incentive mechanisms have continued to be redesigned to reflect ever-changing needs and performance priority areas of NWSC. The other key consideration of incentive design has been compatibility with staff interests and buy-in by top management. After some years, incentive mechanisms became tougher with higher risks taken by Area Managers. Regular performance evaluation meetings gathering all managers, government officials, union representatives, are dedicated to assess the areas' performances. Towards the end of such meetings, trophies, cash awards or sticks/certificates of concern are publically handed over to the best and worst areas based on key performance indicators. In 2013, with the coming of a new MD: Dr Silver Mugisha, evaluation meetings have been re-engineered to include feedback to customers, through annual workshops. During these workshops, top managers present bi-annual results and future targets, and discussions take place with customers and other invited stakeholders like political leaders, the Board of NWSC and Development Partners. In addition, the call centre has been expanded and social media communication platforms opened to provide more outlets for customer service.

The success of IPCs relies on the ability of area managers to identify rooms for improvement, and suggest ways of actually achieving such improvements. NWSC organised a special procedure to select Area Managers and to determine which performance targets and performance improvement schemes are worked on and discussed. The bidding process was introduced to give the chance to any talented and motivated person to bring new ideas and new competencies. The profile of managers has therefore gradually changed. In the 1990s, all managing positions were occupied by engineers. The implementation of IPCs led to greater diversity. Any highly educated person within NWSC, and in Uganda, can bid for the position of Area Manager. So Area Managers may be water engineers, but also may come from the private sector or have an educational background in any field that is easily compatible with sound business management practice. The bidding process makes the performance targets at the area level the result of negotiations between Area Managers and the head office. This approach sets a strong basis for the subsequent accountability of Area Managers.

4.4.2 Key 2: Motivate the staff

The first motivation has been to offer decent salaries (done in the 1990s). Then, with the first IPCs (2000), it was not only the manager who would be praised for the achievements, but the whole areas' staff, with gifts and parties organised along with the trophies ceremony. This has generated a feeling of belonging to the utility and increased emotional incentives. Also, staff expertise has been recognised. Collective design, including staff and managers, to define key targets, have contributed to achieve IPC acceptance by utility staff. Regular meetings gathering staff and managers were organised so that every person could conduct a participatory problem analysis (pointing out strengths, weaknesses, opportunities and threats), identify areas of improvement, and suggest ways of achievement of such improvements. Further improvements in working conditions contributed to enhance the motivation of the staff, with better salaries and regular money rewards, refurbished offices, uniforms, proper tools and SOPs, clarified roles, improved relationships with customers, and a better image.

To avoid free rider behaviours, individual evaluation schemes have been set up to recognise individual performance. The idea is to convince staff that any person in the utility that may have a good idea to improve performance may be given a chance to implement it and therefore may be publicly acknowledged and praised for his/her deeds. This in turn requires managers to know what the motivational drivers are and what is culturally acceptable. At NWSC they chose naming and shaming processes at individual level and Area (group) level to catalyse competition between Areas.

4.4.3 Key 3: Monitor and evaluate performance

It is one thing to introduce IPCs, it is another to implement them. NWSC had to gain the capacity to design contracts, monitor and evaluate performance, and control performance declarations. As a start, NWSC was assisted by experts for the initial designs, mainly for quality assurance. It then developed a special unit dedicated to those tasks, along with soft tools such as contract templates, performance evaluation grids, performance reporting templates and data management systems. From 2004, the head office opened a department called the "Monitoring and Evaluation Unit". This department gathers all performance data sent by Areas to the head office, computes management fees accordingly and manages all conflicts related to the implementation of the contracts. The building of an information system tool was therefore necessary. NWSC chose to use the software SIGMA for a more efficient transfer and storage of information. They also support the selection process of Area Managers. They control areas with the support of operations engineers who make the link between the Monitoring and Evaluation Unit and the areas. Random checking activities are organised to check the integrity and robustness of the reported areas' performance statements and data.

When this unit was created (2004), there used to be 15 areas to benchmark. The unit remained centralised at the head office until NWSC reached 23 operational areas. With the later rapid expansion from 23 to 167 in 2016, monitoring and evaluation has been organised at a regional level. The country has been divided into 4 regions: Kampala, Central, South West and Northern East. In this way, NWSC has developed a unique regulatory system as developed by Mugisha (2011).

This unit plays a key role in regulating the relationships between the head office and areas. For instance, it has to check that targets negotiated between the head office and areas are consistent with what is expected by supervising ministries. Annual performance targets to be reached by the corporation are defined through a process led by a dedicated team at the head office, who works in collaboration with all departments/ sections of the utility (capital investment, operations, commercial, human resource, finance and accounting). They aim at reaching higher performance than the one defined in the Performance Contracts.

4.4.4 Key 4: Manage resistance to change

Change, inevitably, faces resistance. Within a decade, the utility moved from an overcrowded centralised utility with life employment contracts and priorities on making the network work, to a decentralised utility, less staffed but with managers and workers used to working with targets and bonuses, with a commercial focus. This move towards a private-like management style hasn't raised resistance from anti-privatisation campaigners, since it is aimed at preserving the public status of the utility. It is, thanks to the PCs and IPCs' implementation, that the divestiture project was finally abandoned in 2004.

However, the top management faced resistance from staff members with specific resistance from managers due to the potential loss of stability the reform meant to them personally. The reform was based on the idea of a workforce evaluation and use of incentives. The utility had first to downsize so that it could pay less people and create cash margins to pay staff bonuses. Then, life employment contracts were broken in favour of short-term renewable contracts for better flexibility. Everybody had to accept new priorities (commercial orientation rather than emphasis on only engineering matters) and to work with targets and incentives. Whereas decision making used to be centralised, managers were to be given more responsibilities, hence taking the risk of forfeiting bonuses, or even their job. The challenge was huge.

Along with the reform at NWSC, several techniques have been used by the top management to convince managers, staff members and the union that the ongoing transformations would benefit the utility and the staff. Firstly, the top management convinced the (unique) union by making the case that a public utility could be as good or better a performer as a private one that public-sector jobs would be saved by keeping them in the public realm (Muhairwe, 2009).

Once the head of the union was convinced, it was easy for top management to work hand in hand with the union by involving them in every decision involving staff welfare. Until now, a union representative takes part in every welfare-related decision making, evaluation design and implementation process. Conflicts are solved through negotiation processes. For example, the main conflict that was raised in 2005 led to a further increase in salaries of 20% for workers based on an analysis of efficiency gains achieved. Union representatives have in turn been heralds of change in attitudes to work towards discipline, hard work, setting goals and meeting targets. In return, every effort is compensated. For instance, the ending of lifetime employment contracts has been compensated with an increase in salaries and introduction of an annual contract gratuity. Downsizing has been eased by making conditions attractive so that enough staff members become more productive with better salary compensation. This means that a good share of efficiency gains has been invested in human resource management to ensure that NWSC becomes and stays attractive to a qualified workforce.

4.5 IPCs: ONLY CONSTANT IS CHANGE

IPCs must evolve along with the utility. NWSC's top management has not stopped identifying flaws and attempting to improve the design and implementation of IPCs. Following are some key elements that have required some adjustment along the way.

4.5.1 Find the right balance between individual and collective incentive mechanism

According to Muhairwe (2009), it took some years to find the right balance between individual and collective incentive mechanisms. Collective incentives are important to foster teamwork and to generate a sense of belonging to a community of work. It also may lead to free-riding behaviour. That is why the top management also developed individual performance evaluation schemes, based on individual performance agreements (IPAs). When the utility abandoned life employment for renewable work contracts, the idea was to make sure every person employed in the utility gets specific targets and is regularly evaluated. Individual Personal Agreements play a crucial role in the careers of the staff and managers. The IPAs are evaluated bi-annually through a 360 degrees appraisal methodology and staff are given deserving feedback.

4.5.2 Adjust the definition of performance over time

There is no unique way of measuring a utility's performance. The performance depends on the context, the objectives and the specific challenges. When the first IPCs were signed, the main objectives were to focus on financial sustainability. The utility was indeed heavily indebted and had a mission to attract private investors. Therefore, the key performance indicators were linked to revenue generation and financial sustainability. With time, the definition of performance became more complex, integrating for instance, water loss prevention and environmental and social dimensions of performance.

4.5.3 Find the right balance between control and confidence

The success of the implementation of any contract greatly relies on the confidence between both parties. Preserving this confidence requires the ability to dialogue. From the head office's point of view, control is mandatory but risks breaking this confidence. In the benchmarking exercise, there should be a fair evaluation of areas, understanding each area's specificities, without seeming to favour one over another. From the areas' point of view, there may be a fear of being misunderstood. An independent mediating authority is important in such cases. In NWSC, the Managing Director and the Board play the mediation role to ensure fair dealing for both parties.

4.5.4 Keep motivation high

Maintaining the motivation to improve performance becomes harder and harder the higher the utility's performance gets. Consequently, in 2010, it was more difficult to achieve results than it was in 2004. This resulted in a loss of motivation in 2010 because staff had the belief that to work as hard as before would be less likely to achieve any performance bonus. This is a great challenge. Various strategies have been used to motivate the staff. For instance, from 2013, the gap between low and high salaries has been reduced. In addition, the incentive formula has had to be redesigned to reflect changed circumstances making it possible for staff to get a good share of efficiency gains in an equitable manner. For example, the incentive formula now includes reward for preservation of high performance status, compensation for performance improvement (rather than only focussing on targeted achievement) for key performance indicators (KPIs) and cash awards for improving customer satisfaction index (CSI). Mugisha (2011) presents a number of incentive designs that are used in NWSC and other regional utilities in Africa.

4.6 CONCLUSION

IPCs have contributed tremendously to changing the NWSC to make it enter the New Public Management era with a focus on results and customer service. Their successful implementation greatly relies on the foundations that had been built and the specific Ugandan context. Flexibility and adaptability in their management seem to have contributed a lot to their institutionalisation within the utility (Colon, 2014; Colon & Guérin-Schneider, 2015). The zeal to operate "like the private sector" has indeed greatly changed the corporation culture and structure. They have become part of the DNA of the utility. Such an achievement requires an adaptive style of management and the ability to set up learning processes.

Of course, some challenges remain. Four seem worth noting: asset management, focus on sanitation, Areas benchmarking and access to water for the poor. In a country where the demand for water and sanitation services grows at a very high pace, with a small portion of the population able to pay the full cost, area managers tend to prioritise service expansion over management of existing assets. Strengthening incentive mechanisms that favour both service expansion and asset management appear to be a way forward.

Another issue is to find a way to develop both water and sanitation in equitable proportions. The Ugandan utility has not managed to develop the sewerage network as fast as the water one, due to the pressing demand for water and the low availability of funds for sanitation. The utility has only recently launched a large sanitation program and is piloting appropriate low-cost sanitation technologies such as packaged plants in small towns. The objective is to roll out the successful technologies to other towns. Increasing a focus on sanitation will reduce the current pollution of underground water sources through the current extensive use of on-site septic tanks with no proper disposal mechanisms. Clearly, when there is little focus on sanitation, issues of environmental pollution emerge in the long run.

NWSC now operates 167 Areas with a great diversity of situations: from dry to humid regions, from small-scale rural water supply schemes to sophisticated capital city water and sanitation networks. Benchmarking the performance of such a diversity of areas requires development of analytical tools tailored to this complexity. The regionalisation approach is helping in this regard but has not fully eliminated issues relating to disparities. Finally, access to water and sanitation for the poor remains a challenge as in many countries. The utility has developed a pro-poor policy, through a cross-subsidy system embedded in the water and sanitation tariff (the domestic/rich and commercial categories paying more than the low income communal tap-stands), and granting a free connection policy, and is working on pilot projects funded by a number of development partners and the Government of Uganda to expand the services, through pre-paid meter technologies, to the urban poor communities/slums.

This management model may be replicable, provided a number of conditions are fulfilled: the right foundations are in place (working infrastructure, well-trained staff and managers, enabling legal framework) and an organisational culture that is open to the idea of accountability. It requires great motivation and leadership from the Board and top management. The utility, through its external consultancy services department, has worked with a number of regional utilities to improve their performance. These countries include: Kenya, Tanzania, Rwanda, Zambia, Nigeria, Sierra Leone, Ethiopia, South Sudan, India, Bangladesh, Trinadad and Tobaggo, among others.

4.7 NOTE

This chapter is based on two complementary insights on Internal Performance Contracts (IPCs) at NWSC, Uganda: an academic and a managerial one. Dr Colon's PhD thesis provides a deep understanding of the context, the history and the dynamics of the public–public performance contracts while Dr Mugisha's managerial experience adds a practitioner vision.

4.8 REFERENCES

Colon M. (2014). Les contrats de performance dans le secteur de l'eau urbaine ougandais, structures de la matérialisation de la logique de marché et supports du travail institutionnel (Performance Contracts in the Ugandan Urban Water Sector, Structures for the Materialization of Market Logic and Institutional Work Supports). PhD thesis, Institut des Sciences et Industries du Vivant et de l'Environnement (AgroParisTech), Paris.

Colon M. and Guérin-Schneider L. (2015). The reform of New Public Management and the creation of public values: compatible processes? An empirical analysis of public water utilities. *International Review of Administrative Sciences*, **81**(2). Available from: http://ras.sagepub.com/cgi/doi/10.1177/0020852314568837 (accessed 26 January 2015)

Mugisha S. (2011). *Utility Benchmarking and Regulation in Developing Countries: Practical Application of Performance Monitoring and Incentives*. IWA, London, New York.

Muhairwe W. T. (2009). *Making Public Enterprises Work: From Despair to Promise: a Turn Around Account*. IWA Publishers/Fountain Publishers, London, New York, Kampala.

Shirley M. (1998). Why Performance Contracts for State-Owned Enterprises Haven't Worked. *Public Policy for Private Sector – The World Bank*. – Note (150).

Chapter 5

Performance-based service contracts in Navi Mumbai

Vijay Padmanabhan

Indian municipalities must increase the efficiency of their water and sanitation services to meet the needs of their growing populations. The USAID-FIRE (D) project supported the Navi Mumbai Municipal Corporation's (NMMC) efforts to improve the management of these key urban services by introducing consolidated, performance-based service contracts for water supply and sewerage. In 2003, the project was requested to support NMMC combine various service contracts for different parts of the water supply and sewerage systems by assessing their performance, into new performance-based contracts. The objective was to improve service quality by reducing multiple contracts and bring in private sector efficiency. This section describes the design, procurement and monitoring process, and lessons learned about private sector participation in municipal service delivery. It also provides an approach to designing performance-based contracts through a risk assessment where the contracts were designed based on a design-build approach with capital expenditure provided by NMMC based on the bidders' assessment of system improvements. The bidding documents used provided boiler-plates that could eventually be applied for any type of urban basic services subject to a defined set of service delivery specifications and performance indicators. NMMC was successful in engaging contractors for the performance-based contracts and successfully ran them for the initial term of 3 years. There was a mix of national and international contractors that were engaged to deliver the contract.

5.1 OVERVIEW

5.1.1 Context

Navi Mumbai is a planned city on the outskirts of Mumbai, the capital of Maharashtra. As its Marathi name navi implies, it is a new city. The state's City and Industrial Development Corporation (CICDO) planned the city, which was developed in the early 1980s, and in 1998 turned over all major infrastructure

Authors: Jan Janssens, Didier Carron, Philippe Marin, Philip Giantris and Tom Williams
doi: 10.2166/9781780405964_0033

services to the Navi Mumbai Municipal Corporation (NMMC), which had been established in 1991. The city's population in 2003 was 800,000.

The core municipal services were managed by the private sector on a labor contract basis, i.e. contracts based solely on personnel costs, while NMMC provided all material, including equipment and spare parts. Multiple contracts in NMMC covered:

(i) *Water Supply*: (a) operation, repair and maintenance of ground and elevated storage reservoirs, and pump houses within the reservoir premises; (b) repairs and maintenance of internal distribution; and (c) meter reading and issuing bills to customers.
(ii) *Wastewater Services*: (a) Operation, repair and maintenance of sewage pumping stations; (b) operation, repair and maintenance of sewage treatment plants; and (c) repair and maintenance of sewer lines.
(iii) *Solid Waste Management*: (a) street sweeping and drain cleaning; (b) transportation of waste; and (c) maintenance of dumping grounds.

These contracts were renewed annually. The private operators' cost estimates were compared to the Corporation's, expressed as a percentage, and the lowest bidder was awarded the contract. The lack of performance criteria and the large number of contracts made it difficult for the city to control contractor performance. Consequently, the NMMC invited the USAID FIRE project to assess the situation, help them consolidate these contracts, and prepare comprehensive performance-based service contracts (PBC) for each of the three services. The mission of the Indo-US Financial Institutions Reform and Expansion Project – Debt Market Component FIRE(D) Project was to institutionalize the delivery of commercially viable urban environmental infrastructure and services at the local, state and national levels. Since 1994, the Project had been working to support the development of demonstration projects and of a sustainable urban infrastructure finance system.

While this note describes the rationale and obligations involved in the new contract documents for water services, the NMMC followed a similar concept for wastewater and solid waste management.

5.1.2 Current service delivery

In 2003, water was supplied to NMMC from a large reservoir receiving water from the Maharashtra Industrial Development Corporation (MDIC) and the Maharashtra Jeevan Pradhikaran (MJP) water sources. The total supply from these sources was 167 million L/day, nearly the estimated demand, but 30% of the water was lost in transmission and distribution through leaks and being unaccounted for and not billed. Bulk water was supplied to ground level storage reservoirs in each of the city's eight nodes, and then pumped to elevated storage reservoirs where it was disinfected to recommended standards. It was then distributed to customers through a 423-km network of iron and steel pipes (Figure 5.1).

There were 67,663 water connections, metered and flat rate domestic connections and metered commercial and industrial connections. Four private contractors issue monthly bills based on prevailing tariffs while the collection was handled by the NMMC and authorized private banks.

The key issues in service delivery include, but were not limited to, the following:

(i) Service areas assigned to elevated tanks were not proportional to demand and storage capacity, resulting in inequitable supply.
(ii) Connections to feeder mains were subject to pressure fluctuations, resulting in inadequate supply.
(iii) System design originally envisioned sufficient pressure to supply apartments on higher floors of multi-story buildings, but rapid growth, unauthorized connections, and physical losses prevent this objective from being achieved.

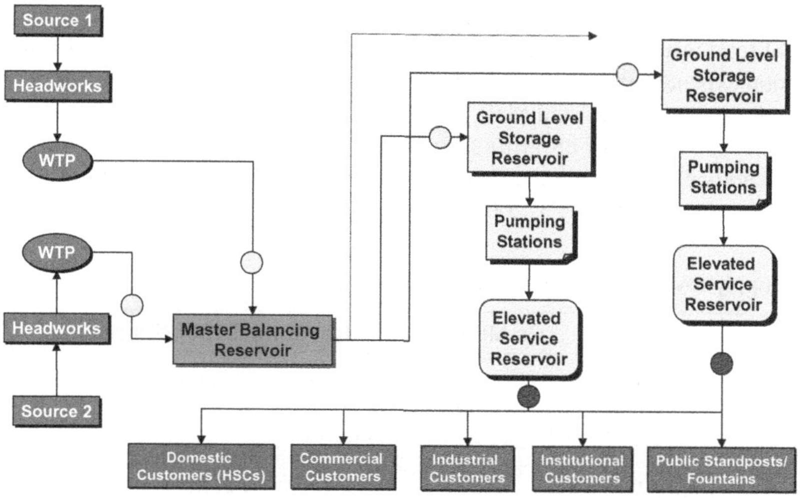

Figure 5.1 Water supply in Navi Mumbai.

5.1.3 Water supply management

The water supply system was operated and maintained by 42 contracts covering billing; operation and maintenance (O&M) of the main reservoir, chlorination plants, pumping stations, and elevated reservoirs; and repairs and maintenance of the feeder mains and distribution systems. These contracts were managed and supervised by the NMMC City Engineer and his staff of four deputy and 14 junior engineers in the Water Supply Department. Figure 5.2 provides the distribution of water supply contracts.

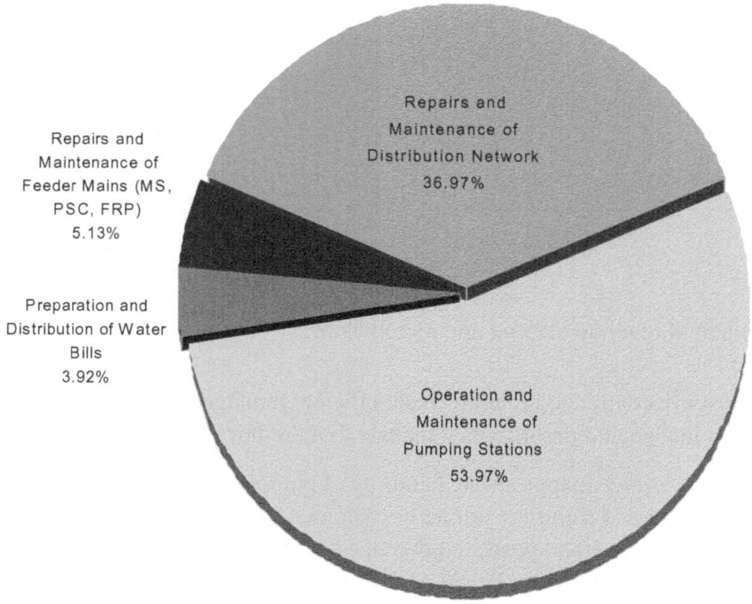

Figure 5.2 Distribution of water supply contracts.

The city's water supply revenue comes principally from water charges received from customers; water tariffs were 87% of total water revenue of Rs 300 million in 2000–01. However, total water revenues were only 60% of the total expenditure on water supply, Rs 503 million. About 75% of the total expenditure was for bulk water purchase from MIDC and MJP. The annual cost of the service contracts was Rs 30 million. Pipeline and pump repairs cost Rs 88 million, or 18% of total water supply expenditure, and included contract services. Another 7% was for electricity charges. Table 5.1 and Figure 5.3 provide a breakdown of the revenue account and statistics for FY 2000–01.

Table 5.1 Water supply revenue income and revenue expenditures, 2000–01.

Revenue Income	Rs Million	Revenue Expenditure	Rs Million
Water charges	259.68	Staff, establishment & admin.	1.94
Connections and others	19.24	Engineering works	0.16
Water benefit tax	21.24	Water purchase	376.62
Income from municipal property	0.30	Pipeline & pump repairs	88.12
		Electrical charges	36.00
Total revenue income	**300.46**	**Total revenue expenditure**	**502.84**
Surplus/deficit	(202.38)		

Source: NMMC, unaudited figures.

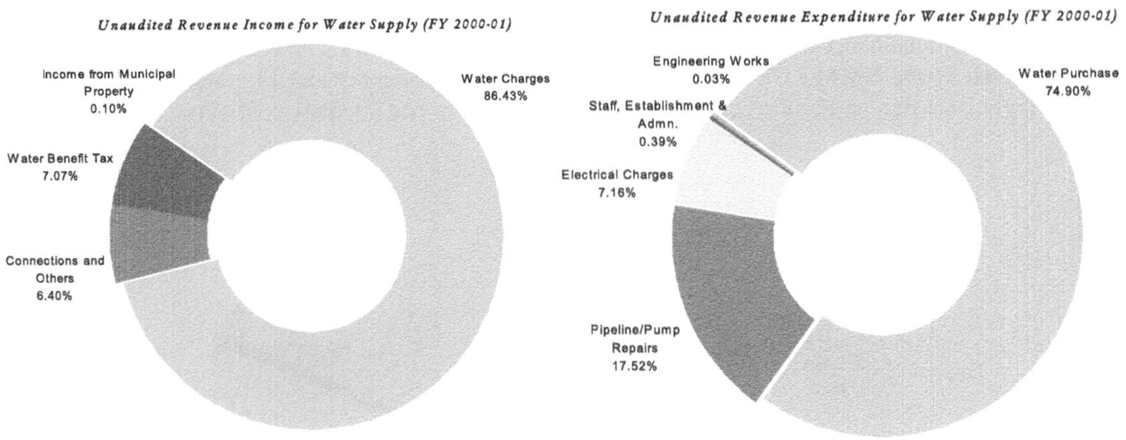

Figure 5.3 Distribution of revenue income and expenditure.

The existing labor-only contracts did not provide efficient management of the city's water supply. The key issues resulting in inadequate performance include, but are not limited to, the following:

(i) *Weak system operations.* Inappropriate handling of pumping led to high energy bills; inappropriate system management and failure to attend to customer queries and complaints about water quality, pressure, leaking public stand posts and damaged pipes resulted in inequitable supply.

(ii) *Lack of water audits.* Failure to detect and rectify leaks and failure to conduct audits to detect illegal use of water, unauthorized connections, and faulty meters led to loss of water and potential revenue.

(iii) *Lack of energy audits.* Failure to reduce energy consumption through regulation of pumping operations and inability to carry out comprehensive maintenance of pumps and/or install efficient pumping equipment to reduce electricity consumption increased expenditure.
(iv) *Lack of contingency plans.* Inability to develop and implement contingency plans responding to natural disasters, periods of water shortage, constraint operations or other similar emergencies.
(v) *Lack of warning systems.* Inability to warn of specific likely future events or circumstances that may adversely affect the operations or conditions of the system.
(vi) *Inappropriate reporting systems.* Inability to provide timely and appropriate information on system condition so as to keep NMMC informed on operations at pumping stations; water supply to customers; water pressure adequacy and equitable supply; water quality; customer complaints and redressal; and customer water meter reading.

5.2 THE PERFORMANCE-BASED CONTRACT DESIGN

5.2.1 Overall nature of the PBC

The existing 42 contracts were repackaged into 18 PBCs to manage water distribution and one PBC to manage the transmission system. NMMC was to handle billing rather than outsource this function; the operator engaged through the PBC will identify customers to be billed and the NMMC will issue bills.

The rationale for repackaging was to improve efficient operation and take specific steps to:

- Maximize the portion of water provided to the transmission system that is billed, striving to reach 100%;
- Reduce, over time, the losses and leakages in the system;
- Detect illegal use of water by customers; and
- Minimize the consumption of electricity.

In the new contracts, the operator was expected to judge the extent of repairs and maintenance required to operate the system for 3 years and accordingly submit its bid. The operator's bid price included administrative charges, consumables, and depreciation on tools, vehicles and equipment, in addition to personnel.

The PBC was designed on an output-based principle with the basic approach that the NMMC would measure the operator on key performance indicators defined in the Services, Incentives and Penalties sections of the Employer's Requirement. NMMC would measure water supplies at input/offtake points and the quality, quantity and pressures at the tail-ends to determine if performance parameters were met within the operator's contract jurisdiction, and accordingly apply incentives and penalties to the contract price/operator fee.

5.2.2 Analysis of the contract structure

Table 5.2 provide key responsibilities of the contractors and Figure 5.4 provides the framework for contracts. The scope of work for these contracts included the following six types of service:

(i) *System Operations* to ensure equitable water distribution, adequate water pressure and quality, proper facility operations, prompt handling of customer queries and complaints, and developing contingency plans.
(ii) *Operations Based on NMMC-Specified Rates* for services such as customer requests made directly to the operator for new connections and sale of potable water.
(iii) *Water Audits* to help operators minimize the amount of water lost in the distribution network and increase revenues. Audits include detecting leakages in the network and at customer connection points and detecting unauthorized/illegal connections and faulty meters. Operators are also required to rectify the problems identified.

Table 5.2 Key roles and responsibilities.

Particulars	Operator	Navi Mumbai Municipal Corporation
Primary task	• Operates and maintains the distribution network inclusive of operation of pumps, elevated and ground storage reservoirs, chlorination, civil works, metering, billing, and collection • Provides all necessary equipment required for operation and maintenance	• Undertakes major capital investment works for the entire water supply system Provides permits and rights of operation to private developer • Oversees operation and maintenance of distribution network
Tariff	• Generates bills as per set tariff	• Determines user charges and tariff structure
Operating expense	• Pays all expenses related to labor employed Bears expenses on repair of assets • Installs meters and incurs operation and maintenance expense on meters • Bears any other operation and maintenance expense related to distribution network	• Pays for raw water purchase, water testing charges, and electricity consumption • Bears expenses on chlorination of water
Capital expense	–	• Bears all capital expenditure works, including augmentation of the existing system, replacement and/or complete rehabilitation of the distribution network
Asset ownership	• Only gets rights of operation of the assets, and is required to hand over the assets at the end of the tenure of the contract	• Takes over assets at the end of the contract period

Source: Toolkit for Public–Private Partnerships in Urban Water Supply for the State of Maharashtra, India (The Government of India – ADB PPP Initiative, 2011).

(iv) *Energy Audits* to help operators minimize power consumption at pumping stations and to follow warranted pumping regulations, maintain power, replace old/worn out pipes, and follow better maintenance practices of electrical installations.

(v) *Repairs and Maintenance*, both preventive and routine maintenance, including minor repairs. The operators will provide spare parts, material, and tools required for routine repairs. However, NMMC will supply material required for certain major breakdowns, at no cost to the operators.

(vi) *Advice to NMMC* on future events or circumstances that may adversely affect operations. The operators also had to advise NMMC on how to improve the quality of operations, reduce water/ energy losses, and improve customer service.

(vii) In addition, contractors are expected to carry out *extra work due to events not defined in the scope of work*, such as extension of service area, at a mutually agreed upon fee.

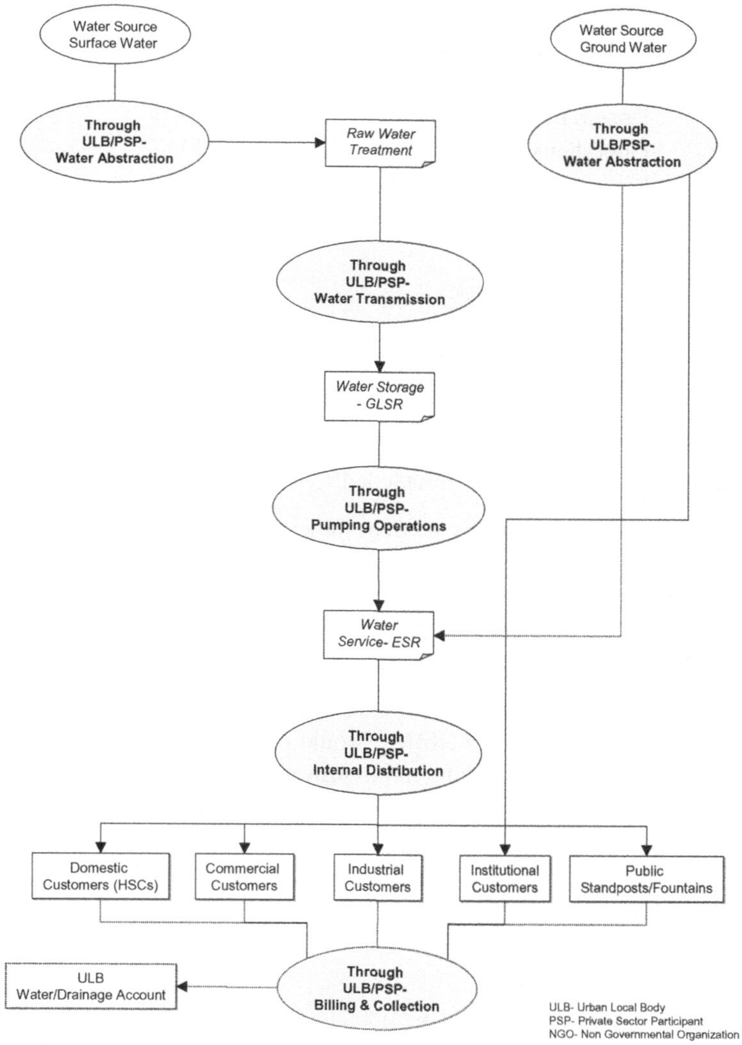

Figure 5.4 Framework for performance-based contracts.

The new contracts envisioned provision of services for 3 years, with an annual performance review required for continuance. NMMC would pay a monthly fee during the first year, with an annual escalation indexed to the wholesale price index.

5.2.3 The procurement process

Eligible bidders included firms short-listed on the Government of Maharashtra's Technical Service Provider list by the Water and Sanitation Department with a minimum of 5 years of experience in operations, repairs, maintenance, and improvement of water supply systems. The risk assessment (Table 5.3) informed the procurement plan. Firms short-listed with other state agencies, including CIDCO, MJP, and MIDC, under specified categories, were also eligible.

Table 5.3 Sharing of key risks.

Risks	Operator	NMMC
Commissioning risk	–	×
Design risk	–	×
Operations risk	×	–
Financial risk	×	–
Payment risk	–	×
Performance risk	×	–
Change in law risk	×	×
Force majeure risk	–	×

Source: Toolkit for Public-Private Partnerships in Urban Water Supply for the State of Maharashtra, India (The Government of India – ADB PPP Initiative, 2011).

Bid evaluation was based on a bid price that included:

(i) The O&M fee for carrying out the six services indicated above for 3 years;
(ii) The power consumption for 3 years – indicated as consumption units (based on the bidder's assessment of pumping equipment performance – existing and new installations); and
(iii) The capital cost of bidder-proposed improvements to the system (based on the bidder's due diligence of the system before submitting bids) – for the purpose of bid evaluation, 60 % of the capital cost of proposed improvements was assumed.

On award of a contract, the operator and NMMC would reconfirm the bidder's assessment of system improvements needed and determine any additional improvement works required for efficient operations, 30 days after the bid is awarded and services commenced. The objective of this exercise was to provide the operator sufficient time to understand the system functioning and verify its technical proposals. The agreed upon improvements would be paid for and carried out by the NMMC and it would pay only the O&M fee component of the total bid price.

5.2.4 Performance monitoring

Benchmarks were set to monitor and evaluate operator performance. Periodic evaluations were to be conducted to determine the percentage of water billed compared to the amount received from the transmission system and the amount of electric power consumed compared to water billed and the State

Electricity Board charges. The benchmarks would be reset following major repairs, replacement, or system improvements.

Operator performance would be rewarded through financial incentives and penalties. Operators will be rewarded for: detection of illegal water use and unauthorized connections; reducing water loss through detection and repair of leaks; and reducing power consumed through optimal pump use. Operators will be penalized if they do not distribute potable water on a fixed schedule and with sufficient pressure. Deviations warranting penalties also include poor water quality, increased power consumption, delays in attending to customer queries and complaints, and provision of unauthorized services.

The incentives and penalties were to be determined every two months and combined before any payments/deductions were made. Calculated on a cumulative basis during an operating year (OY), the following considerations were proposed regarding performance:

(i) In the event that the net accrued penalty at any time of the OY was greater than 10% of the contract price for the OY, NMMC would review contract continuation; and
(ii) In the event that the net accrued incentive payment at any time of OY is greater than 10% of the contract price for the concerned OY, no further incentive payment will be made and the system will be maintained at the level when such incentives cease to apply.

5.3 IMPLEMENTATION AND RESULTS ACHIEVED

5.3.1 Results achieved

The corporation issued 25 new bid packages for water supply and wastewater in March 2003. While 19 new performance-based contracts for water were consolidated from 42 existing contracts, six new contracts for wastewater services were consolidated from 48 existing contracts. The city held a pre-bid meeting to clarify the questions of interested operators and received bids in May 2003. Some firms bid on more than one package. Eight packages in water supply and three in wastewater received non-responsive or single bids and were recalled. The bids were evaluated and those with the lowest price were asked to clarify points in their technical proposals.

NMMC awarded 19 water supply contracts in October 2003 with a total value of Rs 170 million. Four contractors won the 15 main water distribution packages, which were area-based. Bids for maintenance of water supply in two out of the four low-income settlement packages were non-responsive, so the corporation continues to supply water to these areas through labor contracts.

NMMC awarded contracts to operators for three wastewater packages in June 2003 and three other wastewater contracts in October 2003. The annual O&M fee for the wastewater contracts is Rs 100 million.

5.3.2 Sustainability

Per reviews of the PBC conducted by various organizations, during the contract period, the PBC proved to be economical and effective in terms of quality and service of water supply and allowed NMMC to:

(a) Earn indirect revenues through reduction in water wastage, illegal connections, water theft, power bills, breakdown of electrical components and mechanical machinery;
(b) Reduce the maintenance costs of motors and pumps to 65% of initial costs;
(c) Reduce paperwork by computerizing all accounts; and
(d) Achieve 88% collection performance (Administrative Staff College of India, 2007).

NMMC increased its revenue by almost 45% the year following the introduction of the new contracts and reduced unnecessary expenditures over a two-year period; the annual energy consumption was reduced by Rs 4.5 million on sewerage contracts alone. Significantly, customer complaints to NMMC decreased to almost zero. The PBC allowed NMMC not only to provide better service to its customers, but also at lower operational costs (Mulford, 2006).

5.4 LESSONS DRAWN

Key issues raised during the contract design and bidding process include:

(i) The bids were intended to solicit local operators while at the same time incorporating efficiency monitoring mechanisms, but most bids submitted were from large contractors. This indicates that the new service/management contracts were equally favored as large-scale civil construction contracts.
(ii) The requirement for "operator assessment of capital works" was not well understood and so the technical proposals lacked clarity on this point.
(iii) While energy conservation was a key requirement, recommendations by bidders regarding improvements to pumping equipment were ambiguous.
(iv) Another key requirement, detecting illegal connections, was viewed with skepticism. Bidders opined that the activity is politically sensitive and would be difficult to administer.
(v) While the O&M fee for the wastewater service bids were comparable to NMMC estimates, the O&M fee of most water supply bids were substantially higher. This reflected the bidders' apprehensions regarding the expanded scope of work and bid evaluation complexity.
(vi) Bidders did not meet NMMC expectations regarding providing improved service to slum areas, indicating that the city must provide private operators different incentives for these areas.

The main lesson learned has been that the performance-based service contract appears to be a suitable model for private sector participation in municipal service delivery. It allows the urban local body to prioritize investment decisions based on the private operator's assessment of the system. In the mid-term, these 3-year contracts provide the operators sufficient time and exposure to municipal services to bring about system improvements. In the long-term, these contracts provide local bodies the experience and flexibility to move to management contracts and perhaps even long-term concessions with private operators. With the hiring freeze and prudent human resource planning advocated by various states, local bodies could consider adopting this approach for outsourcing service delivery.

The ability to monitor private operators is crucial for the city administrations that engage them. Often, as a result of a freeze on employment, reliance on local private entrepreneurs could have a number of advantages. In the first place, initial experiences show that it leads to financial savings. Second, it compels the board or the corporation to build an information system to monitor the private partner and keep an eye on its cost structure. Furthermore, it could help in building capacity within public utilities as they have to set objectives for the private player, monitor progress, and design adequate contracts. Finally, the emergence of qualified local operators familiar with incentives-based contracts and institutional arrangements is enabled. Subcontracting to small private operators can play a limited role in improving the operations of the sector. Involvement of large private operators, which require other mechanisms for regulation and imply a structural change in the management of the networks are thus not necessarily the most desired solutions for small urban local bodies (Zérah, 2006).

5.5 REFERENCES

Administrative Staff College of India (2007). Public Private Partnerships in Urban Infrastructure and Service Delivery. Hyderabad.

Mulford D. (2006). (US Ambassador to India) India's Water and Sanitation Challenges. The Hindu. Chennai. See http://www.thehindu.com/todays-paper/tp-opinion/article3167540.ece.

The Government of India-ADB PPP Initiative (2011). Mainstreaming PPPs in India – Toolkit for Public–Private Partnerships in Urban Water Supply for the State of Maharashtra, India. Asian Development Bank, Manila.

Zérah M.-H. (2006). Urban water and waste water. In: India Infrastructure Report. Oxford University Press, New Delhi, pp. 130–159.

Chapter 6

Financial comparison of NRW PBCs and conventional NRW projects

Alan Wyatt

This article compares the financial performance of a performance-based contract (PBC) focusing on reducing non-revenue water (NRW) to the conventional approach of NRW reduction and control – usually involving utility staff and engineering consultants/project supervisors. The analysis is based on data from over 100 NRW projects around the world – some PBCs and some conventional projects – which are used to compute project effectiveness and cost efficiency. A general model is used to compare financial performance of the two general options. Results under a variety of local conditions are presented, and general conditions are identified in which a PBC approach is financially superior, and conditions in which the conventional approach is financially superior.

6.1 INTRODUCTION

This article addresses the following issue: from a financial perspective, how does a performance-based contract (PBC) focusing on non-revenue water (NRW) compare to the conventional approach of NRW reduction and control? The author collected data from over 100 NRW projects around the world – some PBCs and some conventional projects (either in-house or in-house with conventional engineering consulting support). The author conducted an analysis of both project effectiveness and cost efficiency and developed a comprehensive framework for financial comparison. This article presents both the methodology and results. In brief, the conclusion is that NRW PBCs are financially advantageous when initial water losses are high, large reductions are needed, and water is expensive; but conventional approaches are advantageous when water losses are low, smaller reductions are needed, and water is less expensive.

6.2 METHODOLOGY

To make the comparison, two types of data were combined in a financial analysis:

(1) the rate of reduction of NRW – also described as project "effectiveness", and
(2) the cost per unit of reduction of NRW – also described as "cost efficiency".

Authors: Jan Janssens, Didier Carron, Philippe Marin, Philip Giantris and Tom Williams
doi: 10.2166/9781780405964_0045

Research was conducted to gather available data on NRW projects around the world from published studies, official documents, and interviews with implementers, on a project-by-project basis. All projects were reviewed carefully for data consistency and plausibility. Roughly half of the projects identified did not have sufficient data of acceptable quality to be useful in an assessment of this type, and were kept out of the analysis.

As will be seen in more detail in the results section below, the PBC projects generally had a distinctly higher project effectiveness and a somewhat lower cost efficiency, when compared to conventional ones. Therefore, the overall financial assessment should be based on a balancing of the effectiveness and cost efficiency. In other words, the comparison issue can be phrased as "is it worth spending additional funds to get faster reduction with a PBC?"

The approach to answering this question is through a financial analysis framework illustrated in Figure 6.1. The concept involves analysis of each of two hypothetical NRW reduction strategies for the same project location (1) rapid reduction (with a PBC) or (2) modest reduction (utility implemented project), in comparison to a passive NRW management approach. The total costs and benefits of each strategy can then be found and compared to each other:

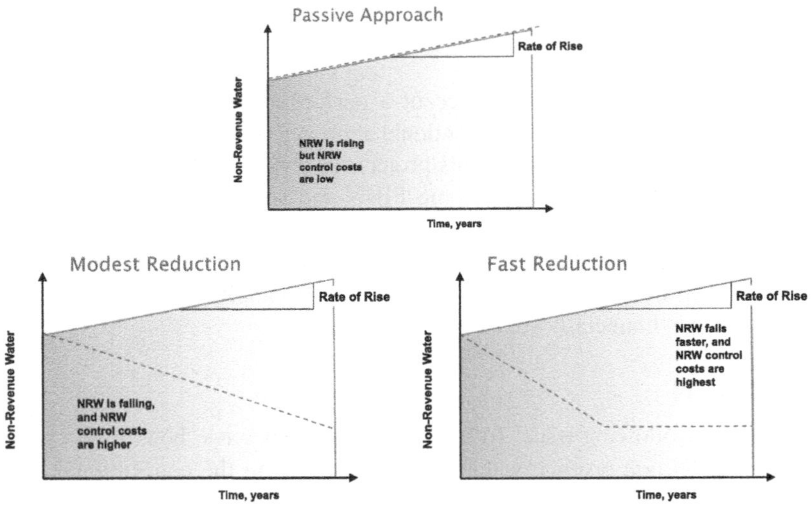

Figure 6.1 Framework for financial analysis of NRW project mechanisms/strategies.

- **Costs**: equal to the present value (PV) of the cost of the NRW reduction and control project (whether slow or fast), less the PV of the cost of the passive program. Since the cost of the passive program is assumed to be zero, the net cost is simply the PV of the cost of the NRW reduction and control project (whether slow or fast).
- **Benefits**: equal to the difference between the total of all NRW related financial values of (1) either the NRW reduction and control project or (2) the passive approach. The NRW related values include: (1) the PV of the cost of the NRW reduction and control project and (2) the PV of the remaining NRW.

In practical terms, the passive approach will have a high value of NRW, but is assumed to have zero NRW reduction and control costs. On the other hand, the reduction projects (whether modest or fast) will have lower value of NRW, but also a significant cost of NRW reduction and control. Furthermore, the modest reduction project has a modest cost and a modest benefit, while the fast reduction project has a higher cost and higher benefits.

A spreadsheet model was used to calculate these costs and benefits using the results of the research on project effectiveness and cost efficiency. The model was used to compute the relative financial performance of PBCs and conventional NRW projects under a wide range of project contexts at two "scales" – a hypothetical location with 50,000 water connections and a second hypothetical location with 500,000 connections – using ranges of values for the following parameters:

- annual growth rate of water supply connections,
- level of NRW at the start of the reduction phase,
- natural rate of rise of NRW,
- rate of NRW reduction under the "project" (from research),
- balance of real losses and apparent losses,
- unit variable cost of water production,
- unit retail cost of water,
- cost efficiency of the NRW reduction "project" (from research),
- and the real discount rate (inflation removed).

6.3 RESULTS

6.3.1 Project effectiveness

The data used to investigate project effectiveness was derived from published reports and official documents on 77 NRW reduction projects in Africa, Asia, Eastern Europe, Latin America, and the Middle East/North Africa Region dating back to 1993, as well as several from the USA and Canada. The dataset includes 23 PBCs and 44 conventional projects. All projects in the database addressed both real losses and apparent losses. Additional PBCs and conventional projects were identified which just reduced one of the two main forms of losses, but those were not included in the analysis. Where sufficient data allowed, NRW reduction programs of greater than 5 years were segmented into phases of 5 years, so that the range of the reduction period across different projects and locations was close to about 5 years in all cases. Figure 6.2 shows results on the project "effectiveness" (rate of reduction of NRW – including estimations of the natural rate of rise of NRW) in relation to the initial NRW level, for NRW projects.

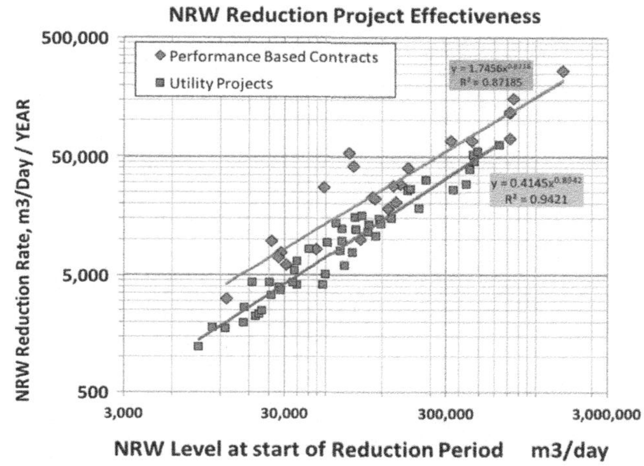

Figure 6.2 NRW project effectiveness.

Overall, as expected, the rate of reduction is higher for projects which start at a higher level of NRW. The two curves show that PBCs are generally able to make more rapid reductions than conventional projects.

Regression analysis shows that the difference between PBCs and conventional projects is statistically significant (>95%). The data scatter is noteworthy – a few conventional projects have reached the reduction rates of PBCs, and vice versa. Also, the difference in reduction rates is smaller at lower initial levels of loss – which seems logical in that utilities reducing losses from a low level probably have some experience with NRW reduction and can be more effective. Utilities starting from a very high level would probably have less experience and not be able to match the effectiveness of experienced PBC contractors.

6.3.2 Project cost efficiency

All projects in the database have effectiveness data, but not all have cost efficiency data. Figure 6.3 shows data on the cost efficiency of NRW reduction projects for projects addressing both real and apparent losses – drawn from locations in the larger dataset for project effectiveness. The results show a very large scatter – reflecting the natural range of costs associated with different projects in different countries. The average cost of conventional projects was US$332 (2010)/m^3/day of reduction, while the PBCs, averaged US$401 (2010)/m^3/day. (Note that the figures for cost efficiency in US $ per m^3/day, are calculated with a recognition that the projects eliminate the rate of rise on NRW and the nominal reduction. There for they are not comparable to other published cost data which does not account for rate of rise.) However, PBCs can achieve better cost efficiency on large scale projects, but are more expensive for small amounts of reduction. While this is a plausible result, and is used in this analysis, it is important to note that given the relatively small dataset, and range of costs in different countries, the cost efficiency distinction is not statistically significant.

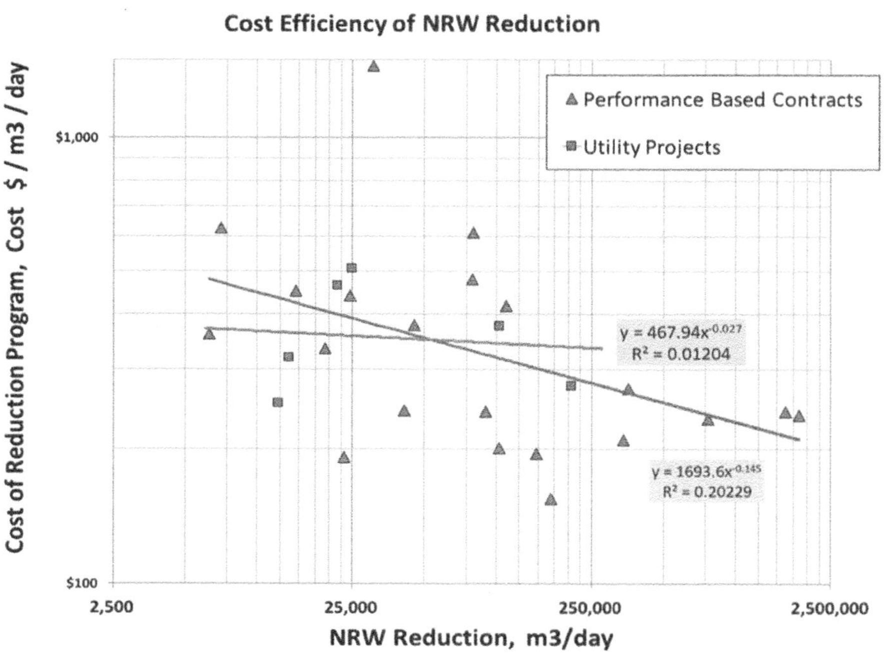

Figure 6.3 NRW project cost efficiency.

6.3.3 Financial comparison

A spreadsheet model was used to generate benefit and cost results for the two types of projects under a range of site conditions. The following parameter values were used to prepare illustrative results (based on typical utility conditions): a growth rate of water connections of 2% per year, a time frame of 10 years, NRW programs structured to reach the optimal NRW level and hold at that level over a maintenance phase, a water price equal to twice the variable water production unit cost, an even split of real losses and apparent losses, and a real discount rate of 10%. Two sets of results were prepared – one for a location of 50,000 water supply connections and another at 500,000 water supply connections.

Table 6.1 shows the results of comparative financial analysis of PBCs and utility implemented NRW projects, based on the method outlined above and the available data on NRW reduction effectiveness and cost efficiency. Specifically, Table 6.1 shows values for benefit/cost ratio using a matrix of inputs (1) a range of initial levels of NRW and (2) a range of variable water production unit costs and retail water prices. The symbols in the table indicate where conventional projects have a higher benefit/cost ratio, where PBCs have a higher benefit/cost ratio and conditions where the two values of benefit/cost are within 10% of each other.

Table 6.1 Financial comparison of NRW PBCs to conventional projects.

Benefit/Cost Ratio for PBCs and Utility Implemented NRW Projects Scale 50,000 connections								
Variable Water Cost ($/m³)	**Low Losses = 300**		**Medium Losses = 600**		**High Losses = 1000**		**Very High Losses = 2000**	
	Utility	**PBC**	**Utility**	**PBC**	**Utility**	**PBC**	**Utility**	**PBC**
0.25	6.4*	4.3	2.7*	2.4	2.2‡	2.3	2.0	2.3†
0.50	10.5*	7.7	6.0‡	5.5	5.3‡	5.4	5.0	5.6†
0.75	13.7*	10.5	9.1‡	8.5	8.3‡	8.5	7.9	8.9†
1.00	16.0*	12.9	12.0‡	11.4	11.2‡	11.5	10.8	12.1†
1.25	19.4*	16.4	14.6‡	14.5	13.6‡	14.6	13.2	15.4†
1.50	20.0*	17.3	17.2‡	16.9	16.6‡	17.3	16.4	18.4†
Benefit/Cost Ratio for PBCs and Utility Implemented NRW Projects Scale = 500,000 Connections								
Variable Water Cost ($/m³)	**Low Losses = 300**		**Medium Losses = 600**		**High Losses = 1000**		**Very High Losses = 2000**	
	Utility	**PBC**	**Utility**	**PBC**	**Utility**	**PBC**	**Utility**	**PBC**
0.25	6.9‡	6.4	2.9	3.7†	2.4	3.5†	2.2	3.6†
0.50	11.2‡	11.2	6.4	8.1†	5.7	8.0†	5.4	8.2†
0.75	14.6‡	15.1	9.7	12.3†	8.9	12.3†	8.5	12.8†
1.00	17.1‡	18.4	12.8	16.3†	11.9	16.4†	11.5	17.3†
1.25	20.7	23.2†	15.5	20.6†	14.5	20.8†	14.0	21.8†
1.50	21.3	24.5†	18.4	23.9†	17.7	24.5†	17.5	26.1†

*Conventional projects have a higher benefit/cost ratio.
†PBCs have a higher benefit/cost ratio.
‡Two values of benefit/cost are within 10% of each other.

Many observations can be made from the results. First, at large scale (bigger cities) PBCs have a financial advantage in all but a few conditions – where losses are low and water prices/costs are low. At a smaller scale, utility implemented projects dominate the "landscape" over PBCs, except in cases of very high losses. Second, at the low initial level of NRW, the B/C values for conventional projects are generally superior to PBCs, especially if water is inexpensive. The difference is smaller as water costs rise. At the high initial levels of NRW, the PBC approach has a superior B/C ratio.

While the difference between B/C values may seem small, that difference involves a difference of millions of dollars of investments and future costs. For example, for the case of 500,000 connections, with an initial NRW of 1000 L/connection/day, and a variable production cost of US$0.50, the utility-led project has a B/C ratio of 5.7 and the PBC project has a B/C ratio of 8.0. The calculation details, from the spreadsheet model, include those in Table 6.2.

Table 6.2 Financial comparison details.

NRW Project Approach	PV of Remaining NRW	PV of NRW Management Program	Total PV	Benefit/ Cost	Program Payback Period, Years	Cost Efficiency, $/m³/day
Passive approach	$1,380 million	NA	$1,380 million	NA	NA	NA
Conventional project	$603 million	$116 million	$719 million	5.7	6	$447
Performance-based contract	$525 million	$96 million	$620 million	8.0	5	$354

It should also be stressed that the B/C ratios can get very high, especially at high water costs – over 20, which is exceptionally high for any infrastructure related project.

BOX 6.1 CONDITIONS UNDER WHICH PBCs ARE EFFECTIVE AND EFFICIENT

1. High level of NRW, especially when combined with low service coverage and/or intermittent water supply
2. Major constraints on water resources, now or soon
3. Large city or number of water connections
4. High water production cost
5. High tariffs (providing good financial return on the reduction of apparent losses)
6. Utility leadership committed to making rapid progress in reducing NRW
7. Political or fiscal pressure for utilities to rapidly reduce water resource withdrawals and rapidly improve NRW, quality of water service, and utility financial condition
8. Limited or low utility expertise in NRW planning and reduction
9. Utility staff, customers and other stakeholders open to the use of private sector expertise to assist the utility

Given the modest amount of input data, especially cost data, these results must be viewed as a qualitative indication of the financial attractiveness of PBC and utility projects, and not the basis for specific project planning.

6.4 CONCLUSIONS

As noted in the Introduction, the principal conclusion from this work is that NRW PBCs are financially advantageous over conventional NRW reduction projects in which initial water losses are high, large reductions are needed, and water is expensive; but conventional approaches are advantageous when water losses are low or small in scale, when smaller reductions are needed, and water is less expensive. The text box below gives an expanded set of conditions in which PBCs would generally be more advantageous in comparison to conventional NRW reduction projects.

Looking forward, the methodology and results do show that careful, ongoing, monitoring of the effectiveness and cost efficiency of all types of NRW projects is very important to develop good guidance on implementation planning and decision making. In addition, new information is currently being collected on locally developed and implemented PBCs in developing countries – which, when analyzed, could provide more useful insights.

Chapter 7

Tegucigalpa PBC case study

David Michaud

This case study describes how the Government of Honduras implemented, with World Bank support, an innovative US$6.5M performance-based contract to improve water service continuity and increase billed volume in a portion of Tegucigalpa, its capital city. Given the limited information available and time constraints, the contract design included three stages (diagnostic, works and maintenance) for a total duration of 30 months. To limit overall risk, the potential performance payments represented only 15% of overall contract amount. Contract implementation took a few months longer than expected, mostly due to delays in the initial diagnostic phase. Performance achievements were solid, with the contractor obtaining around 75% of overall targets and potential payments. However, once the contract ended some of the achievements slid back. Some of the key lessons of the case study include the need to ensure that the design of such contracts is well understood by all parties before being launched; the importance of leaving sufficient time for a solid diagnostic of the issue by the winning contractor; and the necessity to secure the incumbent operator and its staff's full commitment to the contract purpose and design.

7.1 CONTEXT: HOW THE PBC CAME ABOUT

Although a number of performance-based contracts (PBCs) have been implemented worldwide to control water losses, the conditions prevailing in 2007 in the water supply system for the city of Tegucigalpa, the capital city of Honduras, where services are provided by the National Autonomous Water and Sewerage Service (SANAA), were relatively unusual. (This paragraph and the following few are directly based on Garzon, 2011.)

Heavy internal migration had produced a dramatic rise in the population of Tegucigalpa, which increased from 400,000 in 1980 to 1,300,000 in 2008. This growth had not been matched by expansion of the urban infrastructure, and as a result the city suffered from a severe water deficit and a deficient water

Authors: Jan Janssens, Didier Carron, Philippe Marin, Philip Giantris and Tom Williams
doi: 10.2166/9781780405964_0053

network. There was no continuous water service in any part of the city, with the situation worse in the outskirts where water was supplied by tanker trucks. Non-revenue water (NRW) was estimated to amount to around 50%.

Another factor to consider is that much of the data from the system's technical cadaster had disappeared in the wake of Hurricane Mitch, which battered Tegucigalpa in 1998 and caused serious damage not only to the water and sanitation infrastructure, but also the companies' own records. As a result, the cadaster was limited to the main trunk lines.

Finally, SANAA was also operating in an unclear legal framework. A 2003 framework law for the water sector had mandated the decentralization of operational responsibility from SANAA to municipalities by October 2008 (later extended to 2013 and then again postponed) and the transformation of SANAA into a technical agency providing support to the sector's institutions and utility companies. At the same time, signals from the main political stakeholders were unclear with regard to the specific case of Tegucigalpa, where different political parties controlled the various levels of government and unions were very vocal in keeping the status quo.

In 2007, the World Bank committed to provide financing to support the modernization of the water sector in Honduras and promote decentralization of water services to municipal companies, through a project to be called PROMOSAS. Because of the high visibility and complexity of the decentralization in the case of Tegucigalpa, it was agreed that support to Tegucigalpa would be limited to efficiency and service quality improvement activities in a pilot area, rather than a full decentralization.

At that time, there were a number of on-going or planned projects to improve service quality and reduce NRW in Tegucigalpa, all of them based on traditional contracting methods, with SANAA expected to take the leading role in the planning and implementation of the project activities. Based on the successful experience of previous performance-based contracts elsewhere in the World, the World Bank team suggested that activities under PROMOSAS be implemented in the form of a performance-based contract, with payments directly linked with results. The Government and SANAA agreed to explore the concept. (Aside from the activities in support of NRW reduction and service quality improvement in Tegucigalpa, the project included a number of other components aimed at supporting the broader modernization agenda, including decentralization in other cities as well as support to national agencies.)

On that basis, a small team was contracted to advise SANAA and the Government on the design of the contract and to prepare the related bidding documents. A series of informational activities were organized by the World Bank, including video-conferences with leading experts on the topic, and a study tour to Brazil in which several examples of performance-based contracts were reviewed.

7.2 THE DESIGN OF THE PBC

The intermittent nature of the water supply in Tegucigalpa posed major challenges to quantifying water losses and structuring a performance-based contract to control them, since reducing leakages and increasing water continuity represent conflicting objectives, at least initially. Furthermore, the lack of existing information, the limited institutional capacity of the main actors, the relatively short amount of time available for design, bidding and implementation and the desire to demonstrate visible results were all important elements in the design of the contract.

7.2.1 Overall nature of the PBC

Fundamentally, the contract's objective was to demonstrate that quick, visible improvements in service continuity and NRW management could be achieved even in a challenging situation such as Tegucigalpa.

Those improvements were meant to be achieved through better demand management (metering), leakage reduction and an optimization of key elements of the network's structure and operation. Because of the tension described above, decrease in NRW was not an objective per se but a means to improved service (Garzon, 2009).

The contract preparation's main stakeholders were:

- SEFIN, the Honduras Ministry of Finance, as the implementing agency for PROMOSAS and the contracting entity, in charge mainly of fiduciary aspects
- SANAA, the national water utility, as the technical agent and operator of the system, and final beneficiary
- The SANAA employees and union, which had their say in particular on the design of the commercial aspects of the contract (meter installation and reading)
- The World Bank as both the financier and in charge of overseeing the project's technical aspects

One of the main points of discussion during the design, was the extent to which the Contractor's employees would be allowed to conduct commercial activities such as meter installation and reading. Unions were keen to ensure their members would not be pushed out of job, and the final consensus, achieved only during implementation, was to oblige the contractor to be accompanied by SANAA employees for all client-facing activities. The main reason to decide on a PBC was to mitigate the risk of not achieving results despite significant investments. However, the level of risk associated with the contract was deemed as relatively high because of the uncertain Honduras water sector institutional situation and the lack of information on the baseline situation. Early on, the team designing the contract made a decision to settle for a hybrid model where the performance-based remuneration would only be a relatively small part of overall remuneration (15%).

Because of the limited funding available (around US$6.5M) and innovative nature of the PBC, it was decided to focus the contract on a pilot area in Tegucigalpa covering around 40,000 connections, and hydraulically largely independent from the rest of the network. The area represented a diversity of socio-economic and service level conditions.

7.2.2 Analysis of the contract structure

Development of the baseline revealed that only limited information was available, and in the case of the technical cadaster of the pipeline networks, almost none at all. Instead of spending significant time and resources in contracting and defining a comprehensive baseline, the team decided that the Contractor would be tasked with obtaining and processing all the information needed to design an intervention plan, and would be compensated by a high level of flexibility in terms of the activities to be implemented. In that way, the firm would be solely responsible for all phases of the project cycle and would have the necessary flexibility in its contract to bring its technical know-how to bear and propose an optimal course of action.

In order to ensure that prices would be competitively contracted, the bidding document included a preliminary bill of quantity (BoQ) of all expected Goods, Works and Services, including detailed technical specifications for each item. Bidders were expected to base their proposals on this BoQ and the award would be to the lowest qualified bidder (meaning that it would be awarded to the lowest price among all qualified bidders), with the initial diagnostic and performance payments as fixed percentages of the overall contract amount.

The resulting scope and structure of the contract are displayed in Table 7.1.

Table 7.1 Contract scope and structure.

Activity	Phase I Diagnostic 6 Months	Phase II Implementation 18 Months	Phase III Maintenance 6 Months
Component I: Engineering services	**10%,** Lump-sum based		
Component IIA: Works, goods and services for network optimization		**75%,** BoQ/unit prices and quantities-based	
Component IIB: Works, goods and services for technical losses management			
Component IIC: Works, goods and services for commercial losses management			
Component III: Training and knowledge transfer			**0%,** incl. in other BoQ items
Results-based payments for service continuity and metered consumption improvements		**15%,** at months 18 and 30 (each 7.5%), based on performance	

Essentially, the contract's 30-month duration was divided into three phases in the following way:

- **Phase I/Engineering services (6 months):** In this phase the Contractor was tasked with developing a technical and commercial diagnostic of the project area. On the basis of the diagnostic, the Contractor was expected to prepare an "Optimized Works, Goods and Services Plan" (POBSO in Spanish), defining its proposed intervention to achieve the Contract targets. The POBSO's overall amount could not be higher than the Contractor's original Bill of Quantities (BoQ) offer and contractual unit prices had to be respected, but the Contractor was able to propose to shift resources between BoQ items by increasing or decreasing the quantity of specific activities, such as leak detection, meter installation, network renewal, pressure control valves installation, etc. Additional activities not contemplated in the original BoQ could be proposed to a more limited extent.
- **Phase II/Implementation (18 months):** Once the POBSO was approved by the Client (after the supervisor and SANAA revised it), the Contractor could start implementation. Payments were as in any traditional work contract, based on the revised BoQ and the physical progress made by the Contractor. In parallel, at month 18 (after 12 months of implementation) the Contractor was entitled to a first performance-based payment of up to 7.5% of overall contract amount, depending on achievement.
- **Phase III/Maintenance (6 months):** Once all POBSO activities were completed, the Contractor was expected to withdraw most of its team but initiate an intense capacity building and knowledge transfer phase to ensure that the utility could sustain the results achieved. At the end of this six-month period, a second, final performance-based payment of up to 7.5% could be made depending on actual achievements.

The final definition of the scope of the contract required negotiation of a series of agreements between the different stakeholders, SANAA in particular, especially with regard to the management of commercial losses. (This paragraph and the following few are directly based on Garzon 2011.) The task of reading individual meters, for which the contractor was originally intended to have full responsibility, became the subject of a shared arrangement because of concerns raised by the SANAA workers' union (see above). Similarly, while it was planned to turn over the legal prosecution of fraud and clandestine users to the contracted firm, since the objective was to normalize the connection, SANAA wanted to continue handling this process because of its sensitive nature.

Another matter that was the subject of negotiation was the contractor's dedicated access to the SANAA commercial information system and its geographic information system to record all the interventions made during the contract period, such as the identification and legalization of new customers, the installation or replacement of meters, or the meter reading data. Since plans had already been made to update the SANAA commercial management platform during the contract period, it would be incumbent on the contractor to design and implement a process for adapting to the new tool. Furthermore, the contractor would have to provide all the computational tools to ensure effective tracking of the contract achievements in particular for those tasks that it performed that were not included in the SANAA information platform.

In addition, there was discussion of such matters as hours of service, permits to perform interventions on public property, the corporate image to be maintained by the contracted firm, and the social management in which the contractor would have to engage in order to ensure the project's success, in keeping with SANAA policy. It was agreed that the contractor would be responsible for marketing and disseminating information about the project, responding to complaints from users, and handling emergency situations. SANAA reserved the right to implement any community management strategies developed by the contractor in any sector outside the project area.

The final scope of the contract (and original BoQ) included the following activities:

- **Component I/Engineering services:** Update of network cadaster and customer database, evaluation of physical condition of main hydraulic infrastructure, IWA-based water balance and hydraulic modeling.
- **Component IIA/Network optimization:** Installation of District Metering Areas (DMAs), pressure and flow meters, replacement of key network components, installation of pressure control valves and air relief valves, rehabilitation of existing pumping stations and reservoirs.
- **Component IIB/Technical losses management:** Systematic visible and invisible leak detection campaign, repair of detected leaks.
- **Component IIC/Commercial losses management:** Meter reading, installation of meters in unmetered connections, detection and normalization of illegal connections and frauds, analysis and solution to meter reading anomalies and management of small and large diameter meters pool.

The performance-based portion of the contract was deliberately kept simple given the limited background and baseline information available. A total of 15% of overall contract amount was set aside for payment through the following two indicators, to be paid at months 18 (12 months into implementation phase) and 30 (contract closing).

- **Increased service continuity (7.5%):** During contract preparation it was assessed that average service continuity in the pilot area's network was around 4.5 h/day. (Actually 9 h every 2 days, as water was generally distributed every other day.) Based on a preliminary water balance, the design team estimated that commercial and technical loss management measures could increase continuity to around 14 h/day, assuming overall water supply to the service area was not significantly changed.

A safeguard clause was included in case of particularly dry hydrological conditions (based on the raw water produced and measured in SANAA's dam), lowering the targets. Hence the targets were set as presented in Table 7.2, with continuity defined as the time during which water pressure at critical points of the hydraulic network exceeded 1 m of water column (as measured by automated pressure meters at about 20 critical points).

Table 7.2 Contract performance payment scale for continuity indicator.

Performance Payment Percentage	Continuity, h/day	
	Normal Hydrological Conditions	Drought Conditions
0%	4.50	3.00
1.5%	6.50	5.00
6%	12.00	10.00
7.5%	14.00	12.00

- **Increased metered consumption (7.5%):** The contract preparation team established that the pilot area suffered from a high number of unmetered connections, a large number of meters with reading anomalies, and an expected significant number of illegal connections. All three contributed to poor demand management and high commercial losses. The installation of meters was however not uncontroversial in Honduras, in particular in a situation in which service is highly discontinuous. In Honduras as in many countries of Central America, the idea that metering allows for a fairer distribution of water charges is not common; instead it is often seen as a precursor to tariff increases and often associated with commercial practices promoted by the private sector (electricity, communication) and thus a precursor to "privatization". Therefore, strong political commitment coupled with intensive, door-to-door explicative campaigns are necessary to overcome the resistance and explain the rationale for meter installation. It was estimated however that the proportion of metered consumption could be increased by 30% with a solid metering strategy, and the targets were set as presented in Table 7.3.

Table 7.3 Contract performance payment scale for metered consumption indicator.

Performance Payment Percentage	Increase Over Baseline in Metered Consumption
2%	≤ 10%
4%	11–20%
5.6%	21–30%
7.5%	> 30%

The implementation of the Contract was overseen by a dedicated Supervision Consulting Company which was tasked with reviewing the engineering outputs from Component I (including validating the POBSO), supervising that the implementation of the goods, works and services under Component 2 was done in compliance with the POBSO and technical specifications, and monitoring overall performance and approving performance payments.

7.2.3 The procurement/tender process

The tendering process was conducted competitively and took around 13 months from the publication of first call for prequalification, to the award of the contract. The process can only be described as smooth, with no major challenge or complaint received and the final contract price was very close to the estimated cost.

The procurement of the contract was conducted under World Bank rules as the contract was largely funded by PROMOSAS. Because of the overall structure of the project, the bidding process and overall contract management were formally led by the Honduran Ministry of Finance, although SANAA was involved in all stages of the design, evaluation and award process.

A prequalification process was officially launched in May 2010 (SEFIN, 2010a). The main prequalification criteria were in line with usual international practices and included, in addition to strong financial and overall technical experience, past experience in network optimization works, as well as NRW management. During the process, a number of questions were received and several amendments to the prequalification document were issued. A total of six companies or joint-ventures presented prequalification applications and three were finally prequalified (two joint ventures and one company). Both joint-ventures included water operators as well as consulting companies, and the single company had both operational and consulting experience. No complaint was received on the outcome of the prequalification process.

Only prequalified companies were invited to bid and all three submitted an offer; numerous questions were received and the bidding documents were amended on several occasions (SEFIN, 2010b). At bid opening in March 2011, two offers were very close to the estimated contract amount of US$6.5M, while the third was around 20% above (US$7.7M). During the evaluation, two of the three offers were eliminated, one (highest bidder) because of discrepancies between the key staff profile requirements and the proposed candidates, and one because of incorrect bank guarantees. The contract was awarded to the only qualified bidder, which however had also been the one with the lowest evaluated overall price. The contract was finally awarded and signed in June 2011, and no complaint was received on the outcome of the bidding process.

7.3 IMPLEMENTATION AND RESULTS ACHIEVED

As described above, the contract was divided into three main phases: Engineering (6 months); Implementation (18 months) and Transfer (6 months) for a total duration of 30 months. This relatively short duration was dictated by the limits put on the funding source for the contract (World Bank credit). The contract implementation started relatively smoothly, with both the Contractor and the Supervision Company hired in a timely manner.

The Engineering phase started in August 2011 and ended up taking 11 months, mostly because of the challenge in agreeing on the final POBSO, which was finally approved in August 2012. A first version of the POBSO, which had been based on discussions mostly between the Contractor and SANAA was strongly rejected by the World Bank (not an official party to the contract) because it was deviating significantly from the original scope of work and included the development of a SCADA system aimed at providing the basis for better decision making, which however required a significant share of the resources assigned for works to be invested in monitoring equipment instead. After much discussion, a revised POBSO was finally approved in August 2012 and implementation started immediately (some limited, non-controversial activities had started earlier).

The Contractor chose to prioritize activities that could move relatively quickly and provide quick wins. Installation of meters, proceeded very well for example, while works requiring detailed designs were delayed repetitively. In addition, although the works-related activities proceeded smoothly, the service-related activities faced challenges. The leak detection and repair program started but quickly proved futile as no complete water balance had been established and zones could not be prioritized; the DMAs had not been fully implemented and pressure control was not put in place. Contract parties agreed to delay implementation into the subsequent maintenance phase. The reading of meters, which initially should have been done jointly with SANAA meter readers, faced difficulties when the union withdrew its support for the contract's implementation and management declined to enforce that part of the contract in light of upcoming national elections. The exact reasons for the union's change of positions are not fully clear; however, by the time implementation of the contract was in full swing, discussions around the actual decentralization of operations from SANAA to the municipality of Tegucigalpa were becoming more concrete again and might have led to a hardening of the union's negotiating position. First results of the contract were also extremely positive with regards to metering, essentially demonstrating that SANAA staff had not been able to achieve what a private contractor could. As a result, meters were read by the contractor (as required by the contract and for performance payment) but the readings were not incorporated into SANAA's commercial system nor used as the basis for customer billing.

The Implementation phase lasted a few months less than the expected 18 months, but because of the initial delays only finished in October 2013, four months later than expected. In the meantime, the Government of Honduras had obtained a credit extension from the World Bank and the contract was extended to April 2014 to allow for a full 6-month maintenance phase. During the maintenance phase, the original contract stipulated that a comprehensive capacity-building plan be implemented to ensure transfer of competences to SANAA. However, the contractor was late in presenting such a plan and training activities only started in March 2014, as did the postponed leak detection activities, leading to a significant gap between the completion of implementation and the actual training, during which very few activities were taking place.

In terms of overall results, the contract prescribed two results payment deadlines, one at the end of the Implementation phase and the other at the end of the transfer phase, as a way to measure sustainability of the results. The Contractor, Supervising Company and SANAA had previously agreed on a protocol to measure such results (CONHYDRA, 2012), which in part was not fully consistent with the original spirit of the contract, in part because of lack of clarity of the contract's text.

Table 7.4 summarizes the contract progress vis-à-vis the revised indicators and targets.

Table 7.4 Contract indicators, baseline, targets and actual achievements.

				Achievement	
Indicator	**Unit**	**Baseline**	**Target**	**18 Month**	**30 Months**
Increased service continuity	Hours/2 days	8.42[1]	16.84	5.87[2]	13.1
Increased metered consumption	percentage	0%	+30%	+80.29%	+125.26%

[1]This indicator, and its baseline and target were modified to be based on a 2-day cycle given the fact that water is distributed only every other day.
[2]For the first 2 months of this measurement period, the contractor had not installed pressure meters and the Supervision Consultant penalized it by assigning an assumed continuity of 0, which dragged the average down.
Source: CONHYDRA (2012, 2013).

Similarly, the overall contract financial results can be summarized as in Table 7.5.

Both overall results and performance payments remained below the initially expected outcome. In addition, the significant improvement in metered consumption will not be sustained in the short-term, given the fact that SANAA has not incorporated the related meters into its meter reading routes. While the letter of the contract was therefore largely achieved, its spirit was not and the long-term sustainability remains to be seen, largely because of contract implementation issues and to a lesser extent contract design issues. The efficiency of the contract is unclear; in principle the Contractor's own meter readings have shown that SANAA could be recovering about US$600,000 of additional income from the new meters annually, but those are not being read by SANAA; and the limited continuity improvement of around 5 h/day, might not be worthy of the US$50 per capita invested.

Table 7.5 Overall contract value and actual payments.

Concept	Original Contract	Amendments[1]	Actual
Component I/Engineering services	647,838		647,838
Component II/Works, goods and services	4,859,036		4,505,968[2]
Performance bonuses	971,756		736,313[3]
Price increases	NA	+350,000	303,369
Tax adjustments		+145,763	145,763
Total	**6,478,630**	**6,974,393**	**6,339,252**

[1]A total of 8 contract amendments were signed, most of them dealing with administrative or technical aspects of the contract, but also with a time extension as well as an overall price modification to incorporate tax and price adjustment considerations.

[2]The difference stems largely from the leakage detection and repair activities, which were not completed in the scope originally contemplated due to lack of time and disagreement on the methodology. The contractor initially did not conduct any active leak detection activities because it wanted to increase continuity. Once continuity was increased though, the methodology presented was largely ad-hoc and not systematic, and because of the short time remaining, the supervisor recommended to not conduct such activities.

[3]While the increased metered consumption performance bonus was paid in full, only about half of the increased continuity performance bonus was paid.

7.4 LESSONS DRAWN

The main stakeholders involved in the process consider that the contract was neither a dramatic failure, nor a significant success, and generally agree that the general contract design was supportive to achieving the overall objective of improving the quality of service, while challenges at specific design points, as well as during implementation, made it less successful than it could have been. Despite the inability to transform the efficiency gains from the contract into actual monetary benefits to SANAA, its management considers the experience as a success and is planning to reproduce the approach, should another opportunity arise. The contractor and the supervisors were both satisfied with their experience and actually participated in a subsequent bid for a somewhat similar contract in Panama, which is currently expecting to close in early 2017 and has also demonstrated solid results.

Among other things, future similar contracts should incorporate the following lessons:

- Because of their innovative and often complex nature, performance-based contracts can be challenging to fully comprehend, and it is strongly recommended that conceptual design be clearly explained to interested parties (potential contractors and supervisors) during a pre-bid meeting. The organization of an inception workshop with all actors including the contract designers is strongly recommended to help facilitate implementation.
- Given the innovative nature of the contract, the relatively high risk traditionally associated with working in countries such as Honduras and the lack of solid baseline data, the overall

performance-based portion was quite limited (15%). A higher amount would probably have focused the Contractor's attention on achieving improved results.

- The contract's design was based on very limited information; the choice of performance indicators is seen as correct, but their exact measurement, and safeguard conditions could have been defined in greater detail in particular given the high level of uncertainty during design.
- Operator management and staff (including unions)'s full commitment to the contract design is essential and was largely obtained but turned out to waver during implementation. Other mechanisms, possibly legally binding, should be sought to ensure continued coordinated implementation and sustainability of results.
- Given the contract's complexity, the proposed duration of 30 months was too short, and resources assigned, to the engineering phase in particular, were too limited. It is to be seen whether the implementation of such a complex contract and associated supervision challenges, were justified for such a small contract amount, and whether other simpler models might be more appropriate for small contracts.

It is worth mentioning that on the basis of the initial experience of this particular contract, another larger and more developed contract was designed based on a similar general approach to tackle service quality issues in the city of Colon, Panama. This is nearing completion and showing promising results.

7.5 REFERENCES

CONHYDRA (2012). Informe De Medición De Linea Base Y Protocolo Para El Cálculo De La Bonificación Del Contratista UTGP; CONHYDRA; Ministry of Finance, Republic of Honduras, Tegucigalpa.

CONHYDRA (2013). Informe De Cuantificación De La Bonificación Por Desempeño Del Contratista Utgp, Primer Periodo De Evaluación Febrero–Septiembre 2013; CONHYDRA; Ministry of Finance, Republic of Honduras; Tegucigalpa.

Garzon F. (2009). Bases conceptuales; Fabio Garzon; Ministry of Finance, Republic of Honduras, internal report; Tegucigalpa.

Garzon F. (2011). Lessons Learned from the NRW Contract in Tegucigalpa, Honduras; Fabio Garzon; PPIAF/World Bank, internal report; Washington DC.

SEFIN (2010a). Prequalification Document LPI-001-2010 'Servicios de Ingenieria, obras, bienes y servicios para un programa de gestión de perdidas de agua'; Ministry of Finance, Republic of Honduras; Tegucigalpa.

SEFIN (2010b). Bidding Document LPI-001-2010 'Servicios de Ingenieria, obras, bienes y servicios para un programa de gestión de perdidas de agua'; Ministry of Finance, Republic of Honduras; Tegucigalpa.

Chapter 8

Performance-based contracts – key design Issues

Victoria Delmon

Performance-based contracts (PBCs) are used increasingly by utilities to bring about improvements in how water services are delivered. They are contracts to provide specific services or perform a task and differ from traditional contracts for services because at least part of the fee is based on delivery of outputs and outcomes (e.g. reducing non-revenue water (NRW) or leakage by a certain amount). In return, the contractor has flexibility and discretion to determine how to achieve these outcomes. This chapter looks at the main areas covered by PBCs and the key design issues to consider to ensure that the scope of services and incentives are drafted to achieve the improvements in service that the contracting authority is seeking whilst carefully balancing and sharing the risks between the parties. PBCs are evolving and lessons are still being drawn from the early examples. This chapter seeks to draw from these examples whilst noting that these are still work in progress. The World Bank, supported by PPIAF, and in partnership with IWA, is working to simplify and streamline the preparation of NRW PBCs and some of the insights from that exercise are reflected here.

8.1 INTRODUCTION

As discussed elsewhere in this book, performance-based contracts (PBCs) in the water sector are used increasingly by utilities to bring about improvements in how water services are delivered. They are contracts to provide specific services or perform a task and differ from traditional contracts for services because at least part of the contactor remuneration is based on delivery of outputs and outcomes (e.g. reducing non-revenue water (NRW) by a certain amount). In return, the PBC contractor is given flexibility and left to use its expertise to determine how to achieve these outcomes. PBC contracts are typically focused on discrete activities of the utility rather than an outsourcing of all of its functions, although in some cases, the PBC may comprise a more extensive set of activities.

This chapter looks at the main areas covered by PBCs and the key design issues to consider to ensure that the scope of services and incentives are drafted to achieve the improvements in service that the

Authors: Jan Janssens, Didier Carron, Philippe Marin, Philip Giantris and Tom Williams
doi: 10.2166/9781780405964_0063

contracting authority is seeking whilst carefully balancing and sharing the risks between the parties. PBCs are evolving and lessons are still being drawn from the early examples. This chapter seeks to draw from these examples whilst noting that these are still work in progress. The World Bank, supported by PPIAF, and in partnership with IWA, is working to simplify and streamline the preparation of NRW PBCs and some of the insights from that exercise are reflected here.

8.2 ACTIVITIES WHERE PBCs ARE TYPICALLY USED

PBCs have been used for the following activities:

- Customer service centers/complaints handling
- Improving billing and collection (purely commercial) (e.g. in Burkina Faso)
- Leakage reduction only (e.g. in Ho Chi Minh City (HCMC))
- Pressure management in the network (South Africa)
- Meter replacement (San Paolo)
- NRW management (with or without transformation to 24/7 service delivery within the service area)
- Energy efficiency
- Management contract for a part or whole of the service area.

As NRW PBCs have the significant potential to provide multiple benefits to a utility, including improved quality of service, extended asset lives, more efficient operations, reduced greenhouse gases, greater resilience and improved urban competitiveness, this chapter focuses on key design issues related to NRW PBCs.

8.3 EXTERNAL FACTORS THAT CAN INFLUENCE DESIGN OF THE CONTRACT

There are a number of factors that may affect the design of the contract and need to be taken into account to ensure that the contract is sustainable and successful.

Key factors are:

- **Quality of data** – many utilities in developing countries have unreliable or outdated information about their assets and system performance. Addressing these shortcomings can take time and is beyond the scope of a PBC, and thus the contract design needs to accommodate poor quality and incomplete data.
- **Political economy** – if the activity or part of the activity is already being performed by the utility, will the unit assigned to the task be part of the contractor's team and trained up as part of the contract? Will such a unit be willing to take direction from the contractor? Could outsourcing of the activity cause union concern and is there a need for consultation? NRW activities are visible activities – billing and collection by a third party, excavation of roads to repair pipes – are there some activities that the public or politicians would be concerned were being outsourced? If so, are there ways to mitigate this by requiring use of utility uniforms, billing in the name of the utility etc.? In some cases, NRW levels may be linked to collusion or fraud between customers and utility staff, for instance due to meter readings being falsified, or meters being tampered with and slowed. Consideration needs to be given as to whether this is an area that a contractor would be able or willing to manage. For commercial losses, low collection rates can be due to government agencies not paying water bills. Addressing some of these issues may be best managed by a government agency rather than by a private contractor – but, in any case, they need to be carefully identified and the risks to the contractor identified and mitigated.

- **Intermittent versus continuous supply** – a PBC focused on leakage reduction may be more appropriate where the goal is to reduce leakage rather than achieving 24/7 supply. This would be the case where there is already 24/7 pressurized supply, as in HCMC. If the service is intermittent and the utility is reluctant to move to 24/7 supply even within a few service areas, or 24/7 is unlikely to be possible without substantial investment, then improvements achieved under a PBC may be difficult to sustain and a more limited scope may need to be considered. Target setting then becomes important should the PBC aim to reduce volumes of leakage per se or deliver 24/7 – through a range of measures including the reduction of leakage.
- **Availability of competent contractors** – NRW reduction is more than a civil works exercise and requires expertise in pressure management, leakage detection and maintenance of networks. World Bank experience is that this is not an activity that has traditionally been the focus of utilities in many developing countries and so expertise may need to be imported. Bidding selection criteria will need to be carefully crafted to ensure that competent PBC contractors are selected through a rigorous and objective prequalification process (as discussed below). If there is no direct competence in the local market, then the contracting authority will need to be willing to bring in international experience, possibly as part of a joint venture with a national firm. Foreign PBC contractors may not be willing or able to work in a market which is fragile or dangerous. Therefore, the contracting party needs to carry out market sounding early in the design process to determine whether the contractors that they expect to bid would be interested and under what circumstances. Even where there is no expertise in NRW, there may be competency in civil works contracting and so consideration should be given to how to encourage use of the local contractors in these projects to the extent possible.
- **Capacity building component** – to ensure that NRW interventions become more systemic in a utility and in the wider country, it is important to include a specific capacity building component where the utility staff are trained on the job in NRW management. It may be appropriate to be very specific about what this component includes, so that it is not tagged on by the PBC contractor as an afterthought, specifying the number of staff to be trained and the amount of classroom and on the job training that each will receive. This should not be an after-thought component but needs to be central to the contract – without this, any savings achieved will not be sustainable and the discipline will not be institutionalized. Thought should also be given at the outset to the timing for this component. Given that projects often include ambitious targets to achieve NRW reduction, it may be that this component can only be given due attention by the contractor once substantive improvement has been achieved or during the sustaining period.
- **Capacity of the contracting agency to supervise PBCs** – PBCs are more complex than traditional service or construction contracts and the PBC contractors are likely to be more experienced than the utility in NRW. There is likely to be an asymmetry of information and expertise. This can be managed through clear reporting requirements in the contract and by ensuring that there is capacity within the utility through a consultancy which works with and trains up management within the utility to oversee the contractor.

8.4 PARTIES TO THE CONTRACT

For most PBCs, the parties to the contract will be the contracting authority, usually the utility, and the PBC contractor. The contracting authority will have payment obligations to the PBC contractor. There may be instances where it is appropriate to include additional parties to mitigate risks of the asymmetry of information and capacity constraints. In Karnataka, a State level entity was made a party to the contract, alongside the municipal government, to provide assurance to the contractor that capital investment funds would be made available and that the contractor would be paid. The agency also had extensive experience in contract supervision and oversight which the utility did not have.

8.5 SCOPE OF THE CONTRACT AND PBC CONTRACTOR OBLIGATIONS

Determining the scope of the contract and the obligations of the PBC contractor is key to effective contract design. The utility needs to decide what it wants the PBC contractor to deliver and what is feasible in the circumstances, taking into account, where relevant, the external factors listed above. The key point to note from the outset is that under a PBC the contracting authority is looking to the PBC contractor for solutions to a problem that it specifies. Whilst the outputs or outcomes that the contracting authority desires need to be clear, there should be flexibility left in the contract to allow the PBC contractor to determine how best to achieve the targets.

(a) Commercial versus physical losses

Activities covered under *commercial loss management* typically include:

- Surveys to identify missing or unmetered connections
- Repairing or recalibrating meters
- Installing new customer meters
- Identifying illegal connections
- Improving meter reading and data transfer to customer accounting systems
- Improving customer accounting systems
- Confirming customer categorization

Whereas for *physical loss reduction* the typical activities are:

- Establishing district metering areas (DMAs), which isolates water supply to a small sub-set of the utility's customers, improving water inflow and pressure measurement and facilitating leakage detection
- Leakage detection
- Pressure management
- Repairing leaks
- Replacing service connections
- Replacing leaking pipes, valves, hydrants, etc.
- Installing new bulk meters and replacing non-functioning meters
- Identifying illegal connections

Although this distinction is clear in theory, NRW reduction contracts that focus on physical losses may actually include elements of commercial loss reduction where physical losses are measured by the difference between DMA inflows and customer meter readings. The contractor has an inbuilt incentive to ensure meter reading accuracy, to achieve 100% customer meter reading, and will want to regularize illegal connections as these would otherwise be part of the physical leakage calculation. In this case, reducing commercial losses will focus on customer billing and collection. If the scope is to manage physical losses only, then the contract should be focused on leakage reduction and could be measured in terms of the reduction of leakage volumes. As noted below in Section 8.6.3, in all cases, it is important to ensure that perverse incentives are not created by the contract design. PBCs for commercial and physical losses have different cost curves and should ideally be considered separately, even if they are combined into one contract. The challenge here is that often it is not until the project starts being implemented that it is possible to establish the proportion of physical to commercial losses and so the contractual baseline may need to be verified during implementation.

(b) Whole of a utility's service area versus a selected area

NRW and leakage reduction require significant investment and experience shows that sustainable results are only achieved if a systematic approach is taken, creating hydraulically isolated district metering zones where flows into the zone can be accurately measured and customer meters read with accuracy. Given the costs involved, it may be appropriate to select an area of a utility's service area to develop the PBC. If the intention is to move to 24/7, the choice of service area will also depend on whether a continuous supply of water to the area is available and an ability to manage pressure.

Quite apart from the technical constraints on selection of service area, it is often more convenient for a utility to design a PBC for a discrete area to enable the utility to assess the performance of the contractor and benchmark against the rest of the utility. The use of this model allows the utility to decide whether to scale up the use of PBCs or use other means to reduce leakage.

The contract needs to clearly specify where the service area starts (usually at a point where there is a zonal meter to measure the volume of water coming into the area) and where it stops. The contract also needs to be clear whether it covers only existing connections in the area or also includes potential customers. If the intention is to cover all customers, there needs to be a regime built into the contract for new connections and financing thereof (whether fully covered by the customer themselves or subsidized, whether carried out by the utility or the PBC contractor.)

(c) Baseline and target setting for bidding and contractual purposes

In a PBC, a baseline and targets for the particular activity, such as leakage or NRW levels, are needed to determine whether performance has been achieved and they need to be accurately measured. These parameters are needed to allow for consistent bidding and to ensure that the PBC contractor is moving in the right direction in the project. For leakage levels in particular, given that the assets are below ground, it may be difficult to set the baseline and the target with accuracy before contract implementation. The two approaches typically followed are (i) to create pilot DMAs in areas representative of the network as a whole, covering different network characteristics and demographics that may be encountered, from which to extrapolate the metrics for the main project or (ii) to set targets *ex ante* based on best information available but without allowing flexibility in the contract to reset or adjust based on information gathered by the contractor during the contract. In HCMC, the first approach was followed: the leakage target was based on some pilot DMA leak reduction activities (the pilots were selected as being representative of the network as a whole) and the design team worked with the utility to see what reduction was possible in those pilot DMAs and then extrapolated from there. This was the amount put into the bid documents for bidding purposes and some lower amount (50%) was included as minimum performance targets below which contract termination could be considered. Baselines were then prepared one DMA at a time as the project was implemented.

However, in many cases there is limited baseline data and no pilot study to benefit from. Consideration needs to be given then to whether to develop such base data as part of a feasibility study for the project or whether the PBC will include an initial period where the baseline is set once the contractor has been in place and has developed some base data. The latter approach saves time and possibly money and may lead to a more realistic outcome but creates a lack of transparency and creates an opportunity for the contractor to game the situation given that there is likely to be asymmetry of information and a reduced ability to negotiate between the contracting authority and the contractor post-contract award. The bidding process could also be questioned and the contracting authority subjected to complaints from other bidders if the concept of the contract is substantially changed after the award following the outcome of the baseline study.

The approach to addressing data shortfalls is a function of the risks transferred to the contractor. The more risk transferred, the greater is the need to undertake preparatory activities that will provide contractors with sufficient data to allow them to submit bids that provide value for money to the client. It is likely to be in the interests of both parties to have some flexibility built into the contract with a formal adjustment process, even where the baseline and targets have been included in the contract.

When setting targets, it is also important to be realistic about the targets and the time frame for achieving them. Targets that prove too difficult to achieve or that have an unrealistic time frame may be counterproductive, lead to a suboptimal use of resources and lead to an unproductive and confrontational relationship between the parties. They are also likely to trigger a renegotiation.

(d) Leakage reduction versus moving to 24/7 supply

As noted above, if the contracting authority wishes to design an NRW contract where there is currently intermittent supply and low pressure, then targets computed at low pressure and low duration would need to be adjusted to their equivalent under continuous supply and higher pressures.

If the scope of the PBC is also to include increased access to water, then the calculation gets more complicated. In such a case, looking at litre per connection per day per metre pressure may be more appropriate where the system is being transitioned from intermittent to continuous supply, if there is to be an increase in average pressure in the system and if there is an increase in the number of connections to account for the expanding population. This was one of the parameters used in the recent Karnataka contract to move to 24/7 supply.

If the intention is to move to 24/7, the choice of service area will also depend on whether a continuous supply of water to the area is available and an ability to manage pressure. If this is not the case, then the move to 24/7 may not be feasible or may need to be phased.

(e) Increased access to water

This is typically a separate exercise from NRW or leakage reduction but is sometimes included in the same contract, as is the case for the Karnataka 24/7 contract. It raises additional issues if the move is to go from intermittent to continuous supply as the utility will need to ensure that there is sufficient water going into the system, or that the physical loss reductions are sufficient to allow for the likely additional demand. In such a case it may be appropriate to give the PBC contractor full operational management of the system in the relevant service area. The contract will also need to be clear on who finances the connections and when new connections are appropriate (will all new applications be accepted or only those where the cost of connection is reasonable and is not going to have an adverse impact on overall service delivery).

(f) Detection and fixing of leaks

The contracting authority may already have an in-house team or preferred contractor dedicated to fixing leaks and may prefer to maintain this activity. The challenge with this approach, even if the PBC contractor is given oversight of the activity, is that the quality of workmanship or materials used may not be in accordance with best practice and so it may be difficult for the PBC contractor to agree to the sustaining of the leakage or NRW levels. This is a critical area to get right and so the utility should be open to new approaches, even if this is just in a discrete service area. It will also be difficult for the PBC contractor to be judged against achieving and maintaining improvements if it is not in control of the repairs.

Even where the contracting authority agrees that the PBC contractor should control the repairs, there are different models to consider:

- **Bundled model:** PBC contractor is paid per m^3 saved (with a strong incentive to find and fix the biggest leaks first) and will implement the repairs itself (or sub-contract those activities). The benefit of this approach is that the PBC contractor can control the timing of repairs and can apply its own processes to any sub-contracting, and use its own preferred civil works contractors and be in control of the quality of work. The disadvantage for the contracting authority may be that it has preferred local contractors that are familiar with the way that the utility functions and that it trusts. The PBC contractor will also be "at risk" for the upfront financing of the repairs and so if it takes more work than it had predicted to achieve the savings, it may face a significant downside.

- **Unbundled model:** PBC contractor finds leaks and uses public funds through third party contracts to fix them (lower incentive to the PBC contractor to find the biggest leaks first – but incentive to find leaks). This model is lower risk in some ways to the PBC contractor but can lead to profligate use of public money. It may also be difficult for the PBC contractor to control timing of repairs if procurement of the third-party contractor is subject to public procurement processes and possible attendant delays. To increase incentives, the unbundled model could include a capital efficiency gain share or a fixed payment/m^3 saved (computed based on proportion of estimated net present value (NPV) of any savings). There could also be an element, or at least an initial element, that the PBC contractor funds itself so that the timing of the initial work could be assured. A variation of the unbundled model that is common in the UK is separate performance contracts with a leak detection contractor and repair and maintenance contractor. However, in developing countries where the capacity of the contracting authority to manage interfaces between the contractors may be low, this approach might not be suitable.

There could also be hybrid approaches combining these two models, with some activities being implemented by the PBC contractor against an agreed bill of quantities and other costs being included in the performance fee.

The PBC contractor may be willing to finance a portion of the capital works or systems that it is putting in place and recoup the costs through its fee. This will likely have an impact on the portion of the fee that it is willing to accept as performance based, as the project will carry more risk. The financing costs will also need to be factored into early termination payments as the PBC contractor will be concerned to ensure that it recoups its investment in the case of early termination.

8.6 OBLIGATIONS OF THE CONTRACTING AUTHORITY

The obligations of each party need to be clearly set out to reduce the risk of dispute. Key obligations of the contracting authority will include:

- Payment of PBC contractor
- Making the network available to the PBC contractor
- Making water available to the PBC contractor at the delivery points set out in the contract (at quality, quantity and pressure specified in the contract)
- Assisting with permitting (leakage reduction typically requires significant interruption of traffic and the interface with authorities that are responsible for licensing repairs under roads will be key)
- Making funds available for third party contractors where this is not built into the PBC contractor's fee
- Making personnel available when utility staff are to be seconded to the PBC contractor
- Monitoring and oversight
- For foreign contractors, assistance with visas and work permits.

It may also be responsible for implementing repairs, increasing capacity of supply, etc. Failure to meet these obligations will have different impacts on the project and the PBC contractor. The contract needs to specify how this will be managed, where appropriate reducing the obligations of the PBC contractor and, in extreme circumstances, provide for termination of the PBC contract.

There are certain risks that should typically rest with the contracting authority in a PBC. The key one is the regulatory risk associated with the tariff, which is something that the PBC contractor is not in a position to control. It is therefore not appropriate to base a PBC for collections on revenues actually achieved if this assumes certain increases in the tariff which the contractor cannot control. On the other hand, the PBC contractor may need to be protected in the contract against significant tariff increases if these result in lower collection rates due to reduced affordability and willingness to pay.

8.6.1 Monitoring and supervision

As noted in Section 8.4, the utility may not have the experience or capacity to manage the PBC contractor. It may be appropriate for the utility to hire a technical specialist or independent engineer to oversee the work and certify progress against which the performance element of the PBC contractor's fee is paid. This is particularly important in the implementing period when NRW or leakage is being reduced, but is also relevant during the sustaining period. This will also give comfort to the PBC contractor and, in the case of external financing of the required investments, to the funding agency that there is experienced and objective oversight of the project.

8.6.2 Duration and sustainability of contract

In the first generation of PBCs for NRW there was typically a very limited, if any, time period built into the contract when the contractor was required to demonstrate that the savings were sustainable. Whilst achieving reduction in NRW and leakage is a challenge, experience has shown that whether the utility or a contractor carries out the leakage reduction, sustaining the improvements requires on-going systematic management of the system. It is therefore important to build in a significant period during which the contractor is required to sustain the savings into the contract. It is recommended that a sustaining period of at least 2 years is included in the contract. Depending on the complexity of the project and if the PBC contractor is making financial investment, an even longer period may be appropriate. Longer contracts will ensure that the PBC contractor takes a "whole life" cost approach to their work and uses good quality materials and workmanship for repairs.

Whatever the duration of the sustaining period, it is also important that the contract specifies a capacity building component with on-the-job training for the utility staff. This both educates the staff on the approach being taken, with the possibility of replicating it in other zones, and promotes a sustainable approach to network management that can continue once the PBC contractor has left.

It also makes financial sense to ensure that improvements are maintained. As illustrated in Figure 8.1, typically in an NRW, PBC investment will be significant in the early years of the contract, it is over time that that investment is recouped through the efficiency gains in the system. If the savings are maintained they then start to provide a net gain in revenues.

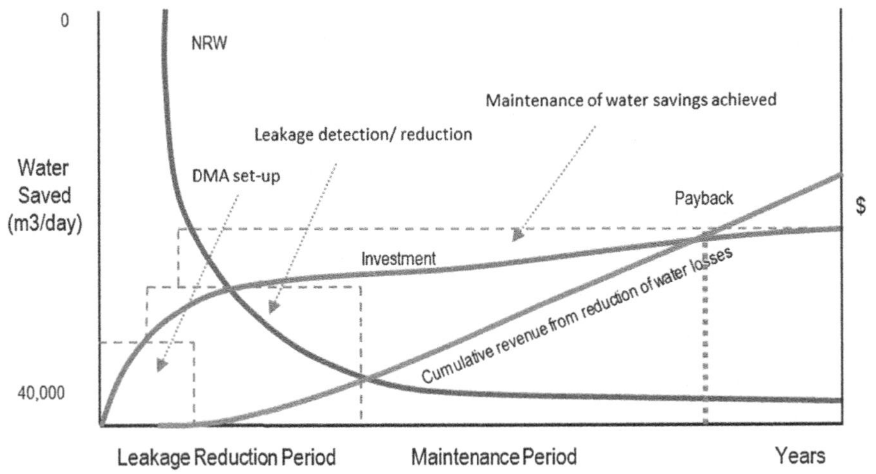

Figure 8.1 Diagram showing the relationship between investment, water saved and increases in investment.
Source: USAID Philippines Water Revolving Fund Report, 2013.

The contract should also make it clear whether and for what maximum duration it may be extended and on what terms – whether it is on the same terms or on terms to be agreed between the parties. An extension of contract period may be limited by local procurement law.

8.6.3 Fees and incentives

Whilst a PBC includes an element of fees that are based on performance of the PBC contractor, a delicate balance needs to be struck in the level of risk that a PBC contractor is willing to, and can realistically take. The proportion of fixed versus performance fee will partly depend on the riskiness of the project and the perceived quality of data at the bidding stage.

If the PBC contractor has an opportunity to verify and rebase the baseline and targets during the initial stages of the contract, then it may be more willing to take a larger element of performance based payment. If performance is dependent on significant contributions from the contracting authority, such as repair of leaks identified by the PBC contractor, the PBC contractor may not be confident that it can control the outcome and so be less willing to have a substantial portion of its fee based on performance. If the PBC contractor is implementing the works itself then it is going to be in greater control of the quality of the works and be building into its price a margin for project management and construction. It may then be willing to have a larger percentage of its fee linked to performance.

Based on feedback from potential bidders, it is clear that their preference is for a PBC contractor to recover most of its costs under the fixed element of the fee and typically would be willing to have some or all of the profit element at risk. Expecting significantly more than this may be unrealistic but may ultimately depend on the quality of information available to bidders and any experience from existing PBCs or pilot schemes. However, the contracting authority will also want to ensure that there are real incentives built into the contract to maximize performance, as too minimal a performance element may provide insufficient incentives to perform. The performance portion of the fee should therefore be carefully chosen based on a balance of data quality, likelihood of achieving the targets and the upside/downside of over/under performance.

It is recommended to make the performance payment dependent on cumulative improvements so that the PBC contractor is incentivized to sustain and build on earlier improvements.

As noted above, under "Detection and fixing of leaks", the contracting authority may be responsible for financing all of the repairs and capital investment, in which case there should be a transparent mechanism for setting costs, such as a bill of quantities in the contract, if the PBC contractor is to be responsible for implementing the repairs, or a clear mechanism for third party contracting.

Damages/penalties: whilst most of the incentives for performance will be built into the performance fee, the contracting authority will need to consider whether there are specific areas where it needs to impose damages/penalties for non-performance or delay in meeting targets. Whilst damages should be used sparingly to keep the focus on incentives, and should not lead to "double counting" of damages, they can be useful if there are specific areas of performance which the contracting authority prioritizes and which may not be covered by the performance incentives. In most common law countries, it is important that any pre-defined damages reflect reasonable estimate of loss and should not be seen as penalizing the PBC contractor, otherwise such provisions could be deemed void by the courts.

8.6.4 Dispute resolution

As noted above, there is potential for dispute and renegotiation of the baseline and targets in even the best drafted PBCs, as it is difficult to have full accuracy of baseline data. The contract should therefore have a mechanism for rebasing the baseline and targets beyond certain parameters, or an initial period for setting the baseline. Given that this is a highly technical area, it is recommended that a technical expert, such as an independent engineer, be employed to verify data, the need for adjustments and capital expenditure. The parties will need to specify in the contract (assuming that the legal system allows for this) whether decisions of the independent engineer should be binding or they can be subject to dispute resolution.

Even where there is flexibility built into the contract to adjust baseline and data, differences are likely to arise between the parties from time to time, so the contract should be designed to anticipate and manage those disputes before they escalate. The contract should provide for a mechanism for reporting and regular discussion of issues at a technical level, with each party appointing representatives for this purpose. Experience shows that where good communication exists between the parties, then issues tend to get resolved before they escalate.

The contract should specify the forum for resolving disputes that do escalate. A mechanism for mediating a dispute, using a third party to help the parties find an amicable solution is recommended. Professional national or international mediators can be selected by the parties to carry out this role, and the mechanism for selection of the mediator should be set out in the contract.

Consideration should also be given to whether resolution through the courts or arbitration is appropriate. Whilst most PBC contracts are not of such significant value that resort to international arbitration may be appropriate, local arbitration may be a cost-effective and more efficient mechanism for dispute resolution than the courts in jurisdictions where courts have limited capacity to decide issues of a technical nature and/or where they are operating under a significant backlog. For larger, more risky projects, international arbitration may be more appropriate.

8.6.5 Early termination

The contract should set out the conditions for early termination and if and when either party may trigger early termination. The consequences of termination should also be spelt out, in particular whether there is compensation payable to the PBC contractor. Where the PBC contractor finances any of the works or systems it would typically be compensated for this even if the termination is due to its default. In the case of termination for contracting authority default, there should be an element of compensation for lost revenues to the PBC contractor and even for loss of profit. At least, there should be payment for the cost of retrenchment by the PBC contractor.

8.6.6 Procurement

As noted above, thought needs to be given to the bidding parameters for the project and whether international or local bidding is used. Local law should also be checked to ensure that the contracting authority has the right to outsource the activity being proposed under the PBC contract.

Whether to follow local or international procurement will partly depend on the size of the project but should also take into account whether relevant expertise is available in the country. It may be that there is relevant expertise in respect of leakage repair, but not in respect of NRW management or leakage detection, in which case the tender documents may seek to encourage local subcontracting, and/or the use of joint ventures, subject to local procurement rules.

As NRW management is a skilled activity which is not extensively practiced in developing countries, the contracting authority will want to ensure the technical quality of the bidders and bids. This could be through the technical bid but it may be considered more appropriate to include a prequalification process to allow only suitably qualified contractors to make bids. The prequalification process should identify the key skillset required from potential bidders and the scale of such skillsets in relation to the size of the project being bid, identifying both the technical and financial capacity that bidders (including consortia of bidders) should have.

Parameters of the technical bid need to be considered carefully and should concentrate on those elements that are crucial to the success of the project. The contracting authority will also need to consider whether there should be a pass/fail for technical bids or a weighted scoring applied and how different technical bids will be compared.

Financial bids – the contacting authority needs to consider what the basis for evaluating financial bids should be. Will bidders quote against a bill of quantities for fixed costs/or should they bid a fee for activities, or a combination of these? As some activities will be being completed over extended periods, the bidding documents need to specify how bid prices will be compared such as through use of discounting to determine the net present value of each bid, and for inflation. Where there are materials that are likely to need to be imported, the bidding documents also need to indicate what currencies the bids can be made in and if in foreign currency, how that will be converted into local currencies for bid evaluation.

PBCs involve a mixture of knowledge services (core to the success of NRW PBC) and construction activities. As noted in Section 8.6.3, the contracting authority will need to give thought on how to ensure that the project does not become more of a construction contract, where the incentives and the bulk of the payment is focused on civil works. The unbundled model mitigates this risk to some extent. Where the bundled model is preferred, as in the case of HCMC, the construction activities were limited to be no more than 50% of the total bid price, thus forcing bidders to push capital costs into the knowledge element of the project.

8.7 ACKNOWLEDGEMENTS

This chapter was written with insights from World Bank colleagues with far more experience than me in the technicalities of NRW including Bill Kingdom, Charles Delfieux, Jemma Sy, Gerard Soppe, Philippe Marin and Rajesh Balasubramanian.

8.8 OTHER RESOURCES

As noted already the World Bank supported by PPIAF and in collaboration with IWA is developing tools to simplify and streamline PBCs for NRW. In addition, the following resources can be useful:

- PPP in Infrastructure Resource Center – ppp.worldbank.org which includes sample PBC contracts as well as other materials on the legal and commercial structuring of PPPs and similar contracts.
- WSP Toolkit for structuring PSP contracts for small scale water projects (2014) – See http://www.wsp.org/sites/wsp.org/files/publications/PPP-Contracting-Toolkit.pdf
- Kingdom, W., Liemberger, R. and Marin, P. (2006). The challenge of reducing non-revenue water in developing countries, how the Private Sector can help: A look at performance based service contracting. World Bank.
- International Benchmarking Network for NRW industry benchmarks (ib-net.org).

Chapter 9

NRW reduction optimization framework

David Ehrhardt

Non-revenue water should be reduced to optimal levels. The optimal level is the level where the long run economic cost of reducing NRW by one unit equals the value of the NRW saved. The marginal cost of reducing NRW can be estimated from an efficiency NRW reduction plan. The economic benefit of physical loss reduction depends on whether water is intermittently supplied or continuous, as well as the cost of additional water resource. Reducing commercial losses has mainly financial benefits which depend on the amount of additional cash collected. The optimal levels of NRW can be used to set targets for performance-based contracts for NRW reduction.

9.1 INTRODUCTION

This paper sets out an analytic approach to estimating the optimal non-revenue water (NRW) level, and outlines how this framework can be used in designing performance-based contracts (PBCs) for NRW reduction. NRW reduction programs all contain targets for NRW reduction, and budgets for the achievement of those targets. In PBCs, those targets are frequently incorporated into the contract, with the contractor rewarded for meeting them, or penalized for falling short. Incentive payments are set based on estimates of the costs of reducing NRW, or the value of the reductions achieved. All this means that, in designing NRW reduction programs and contracts, we need to know what level of target to aim for. We also need to know how much we should be willing to pay to achieve any particular goal. Figuring out these levels and payments requires an optimization approach, as outlined in this paper.

A common practice is to set targets at some assumed "best practice" level. This common practice is not best practice. The "right" target for loss reduction is always situation specific. Where water resources are plentiful and cheap, we should spend less on NRW reduction than where water is very scarce and expensive to provide.

Authors: Jan Janssens, Didier Carron, Philippe Marin, Philip Giantris and Tom Williams
doi: 10.2166/9781780405964_0075

This paper offers an optimization framework that provides the basis for setting targets and payments in a rational way that will deliver the greatest value for the public. The paper starts by describing the basic optimization concept common in the literature, the Economic Level of Leakage (Section 9.2). Section 9.3 then looks at how optimization would change in conditions of intermittent supply. Optimizing commercial loss reduction is then added (Section 9.4). These concepts, along with possible gains in commercial efficiency and energy efficiency, are then combined into a general optimization framework in Section 9.5. Key questions for discussion are presented in Section 9.6. The paper then shows how these concepts can be applied in optimal contract design.

9.2 OPTIMIZING PHYSICAL LOSSES

Reduction in physical losses should be pursued until such point that the cost of saving an extra unit of water is equal to the value of that unit of water. Below, we discuss this optimization concept, and how both the costs and benefits of reducing physical losses can be estimated.

9.2.1 The point at which physical losses are optimized

The optimal quantity of losses is that quantity which maximizes net economic benefits. This quantity is found at the point where spending an additional dollar for loss reduction would yield exactly one dollar in benefit. This is because:

- Different loss reduction interventions have differing costs per unit saved. This is illustrated in the "Leakage Reduction Supply Curve" in Figure 9.1 (This is often referred to as the cost curve. However, this paper uses the term "supply curve" because this represents the supply of loss reduction which a utility may buy through a performance based loss reduction contract. The 'value of a m^3' line then shows what the demand curve of a rational utility would be.)
- The water saved has a monetary value. This is shown as the "value of a cubic metre leak reduction" line in Figure 9.1 (a horizontal line, for reasons explained shortly)
- It makes sense to take all loss reduction steps that cost less per unit saved than the value of the unit. However, it makes no sense to do interventions that cost more than they deliver in value.

As Figure 9.1 illustrates, the optimal level of NRW is found at the point that the NRW Reduction Supply Curve intersects the "Value of NRW Reduction" line. This is marked as point "O" on the diagram. (If NRW were below this point (say at point N) it would make sense to implement the remaining NRW reduction activities from point N to point O – all these interventions cost less than the value they produce. On the other hand, activities to reduce NRW even more (say to point P) should not be undertaken, as they would cost more than the value they deliver.) Net benefits of loss reduction at this point may be represented by the green shaded area. This area is the difference between total benefits (the entire striped rectangle) and total costs of the program (the blue striped area).

In reading the diagram, keep in mind that the horizontal axis shows quantity, not time. It can be misleading to read the figure as showing interventions on the left as being done first, or that cash expenditures will start low and then rise. On the contrary, all interventions with a unit cost lower than the value of water saved should be implemented—possibly simultaneously. Some of these interventions will have significant up-front costs (for example, creation of district metering areas), and only save water over time and in conjunction with other expenditures, such as active leak detection and repair.

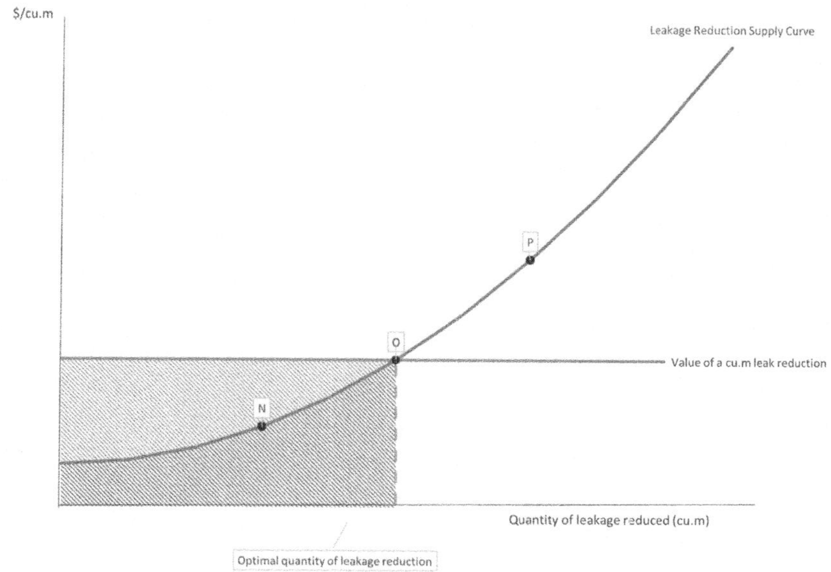

Figure 9.1 NRW optimization concept.
Note: The costs and benefits referred to are both measured over the life of the project. The cost includes all costs needed to achieve the leakage reduction. If a leakage reduction intervention requires both an upfront capital investment, such as installing district meters, and cost spent on fixing leaks, the total cost of the program (in present value terms) divided by the amount of water saved would give the unit cost of that intervention. This supply curve is made up of several such interventions, as illustrated in Figure 9.2.

9.2.2 Deriving the supply curve for physical loss reduction

As Figure 9.2 illustrates, the leakage reduction supply curve is compiled from the cost curves of several physical loss control interventions. The curve starts with the cost of the cheapest option. After diminishing marginal returns means that the cost of the cheapest option starts to exceed the cost of the next cheapest option, the next option is shown, and so on.

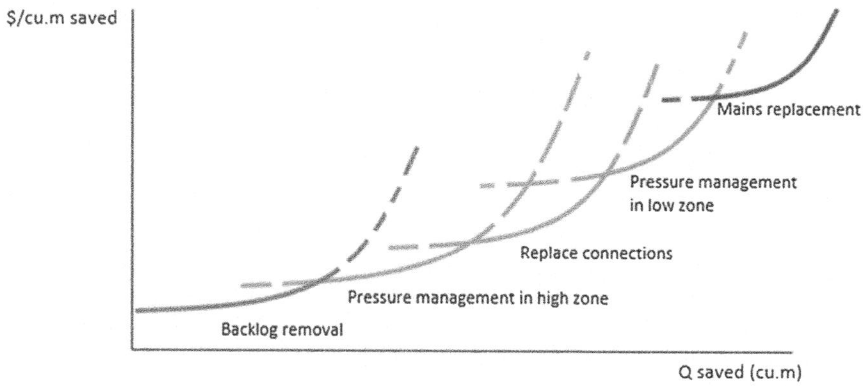

Figure 9.2 Illustrative composition of the leakage reduction supply curve.

9.2.3 The various benefits of physical loss reduction

In a situation of continuous supply, reducing physical losses brings three main benefits:

1. Reducing variable operating costs, primarily the electricity used in pumping, and chemicals used in treatment. These costs are known as the short run marginal cost of water (SRMC)
2. Deferring the need for capital expenditure to increase production to meet growing demand (assuming that demand is growing). These capital costs, when added to savings in variable operating costs (SRMC), are known as the long run marginal cost of water (LRMC)
3. Reducing the quantity of water resource abstracted from the environment, thus leaving more resource available for other abstractors (such as agriculture) or environmental uses (Tripartite Group, 2002). If this value of the water resource is added to the LRMC as defined above, we may term it the "Social LRMC".

These various benefits are shown in Figure 9.3 below.

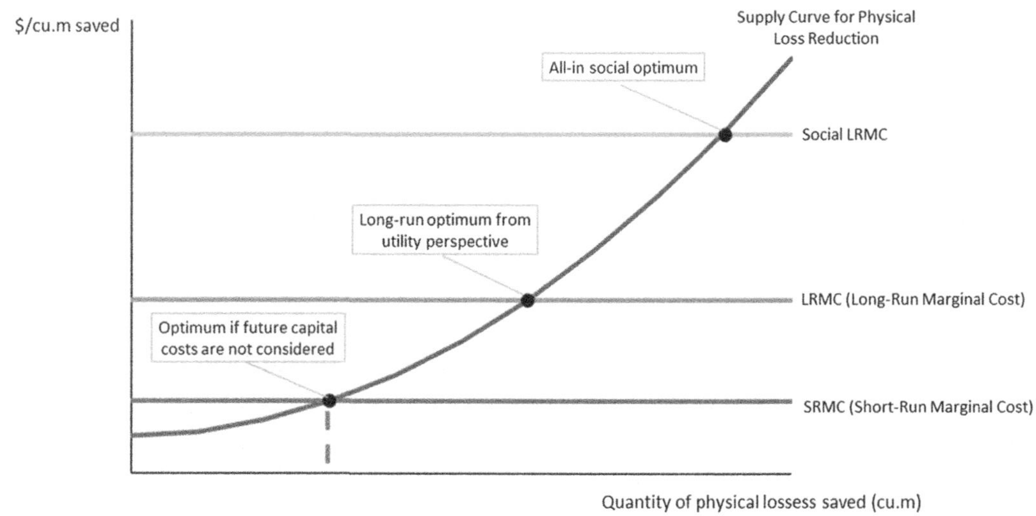

Figure 9.3 Various benefits of physical loss reduction create various optima.

As Figure 9.3 shows, the optimal level of physical loss reduction depends on which of these values of physical water is considered. Utilities will clearly save the SRMC, and so should be willing to pay for that level of savings.

Utilities should also be willing to pay for savings in LRMC; Consider that utilities need to serve an expanding population. To serve more people, they need to invest in additional water production. Loss reduction programs reduce the investment in water production infrastructure needed. Utilities should be willing to pay a dollar for loss reduction if it reduces the water production expansion costs by a dollar or more.

Funding for production expansion programs may come from a funding entity other than the utility, such as a ministry of finance, or a municipal government. The funding entity should be willing to fund loss reduction if by doing so it reduces the production expansion investment needed by a greater amount.

Where water resources are scarce, the utility may need to factor in the value of the resource in alternative uses. This could be the case if, for example, the utility has to buy water rights from farmers, or pay an abstraction charge set by a water resource regulator. In such cases, it will be optimal to value a reduction in physical losses at the social LRMC. This approach is in line with the "Avoided Cost Model" for valuing physical loss reduction adopted in California (Dickinson, 2007).

BOX 9.1 WHY NOT THE TARIFF AS THE VALUE?

The astute reader will have noted that we have not proposed the tariff as a possible measure of value. Why not? If there is unmet demand, water saved by a reduction in physical losses could go into extra sales.

Their answer to "why not?" is that (i) we are concerned with economic, not financial, costs and benefits, and (ii) we assume that the utility must ultimately meet all the demand in the city. It is true that physical loss reduction can lead to increased sales. However, these sales should be valued at their economic value, not their financial value. The economic value is not the tariff, but the value of the water to people. This must be higher than the tariff, or the water could not be sold.

So, should the water saved be valued at the benefit it brings to people? No, because bulk supply increments can also bring water to people. In growing cities, bulk supply increments will ultimately be required, and these set the marginal value of water. This is why LRMC, not the tariff, is the value to use for long-run optimization. Ultimately, everyone needs to be served, and serving them will require investment in new bulk supply. Physical loss reduction defers the need to invest in bulk supply, and so its value is the saving in production investment. Setting the value of physical loss reduction at the tariff will lead to under-investment in loss reduction, and ultimately to a higher cost of service to consumers.

Of course, it may be argued that setting the value of loss reduction at the tariff helps to ensure that the program is financially positive for the utility. This is true. But in the long run, it is wasteful for society as a whole. A good option is often to involve the Ministry of Finance (which usually has to fund the new resource investment) and demonstrate how funding physical loss reduction will ultimately cost less than funding new resource expansion.

9.2.4 Estimating the value of deferred capital costs

To estimate LRMC, we need to estimate the cost of expanding water production to meet growing demand. As Figure 9.4 illustrates, utilities need to add bulk production in increments as demand grows. The optimal approach is generally to add new schemes in order of increasing cost per MLD. Utilities that want to achieve reliable supply will ensure they have a certain reserve margin above peak demand, and will try to bring on new capacity just in time to never fall below this reserve margin. This is illustrated at the top panel of Figure 9.4.

A reduction in physical losses is illustrated in the bottom panel as a "one-off sustained drop in demand". This refers to demand for bulk water input here. Physical loss reduction will create a sustained reduction in the physical losses, and so in water input needed. With this lower demand, the next capacity expansion can be deferred – shown by the pushing back of all the supply increments.

Because of the time value of money, deferring capital expenditure has real value. This value can be calculated by the difference in present value of the capital expenditure program for capacity expansion as originally planned, and the present value of the deferred capacity expansion expenditure. This concept can be expressed in the following formula.

$$\text{LRMC of Capital}\left(\frac{\$}{MLD}\right) = \$PV[\text{Expansion plan 1}] - \$PV[\text{Expansion Plan 2}]/$$

$$[\text{One} - \text{off sustained drop in demand } MLD]$$

Where:

- *PV* = the present value
- *Expansion plan 1* = The investment in production needed with absent leakage reduction
- *Expansion plan 2* = The investment in production needed if leakage reduction takes place, creating a one-off sustained drop in demand
- *One-off sustained drop in demand* = The reduction in demand for bulk water created by the reduction in leakage brought about by a leakage reduction program
- *MLD* = Million liters per day
- $ = dollars.

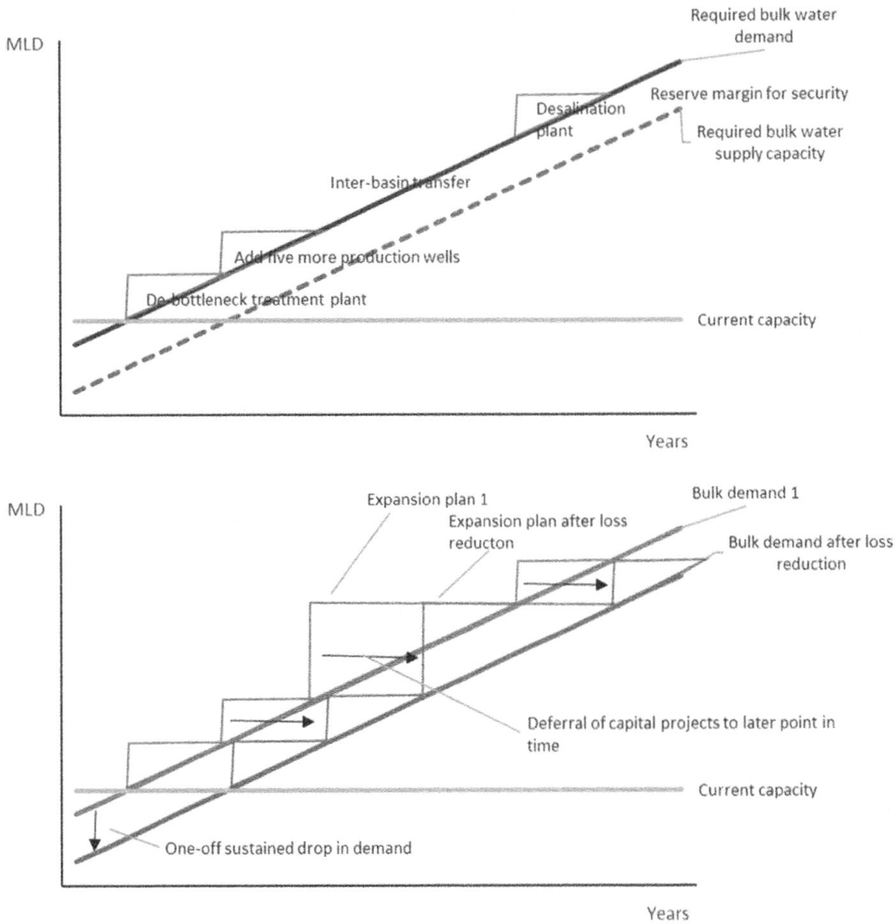

Figure 9.4 Value of deferring the capital expansion plan.

Alternatively, one can use the long run average incremental cost measure (LRAIC).

$$\text{LRAIC}\left(\frac{\$}{\text{m}^3}\right) = PV[\textit{next scheme}]/PV[\text{total output from next scheme over its life (m}^3)]$$

Where:

- *PV[next scheme]* = The present value in dollar terms of the costs of the water production project the utility plans to implement next
- m^3 = Cubic metres of water.

9.3 OPTIMIZING FOR SERVICE

In conditions of intermittent supply, interventions which reduce the leakiness of the network will, generally, be used to increase hours of supply, not to reduce bulk water input. In these situations, the benefit of physical loss reduction will be extended hours of supply.

This section describes how hours of supply and physical losses are interlinked. It then offers a framework for valuing increased duration of supply, estimating the costs of bringing a household onto continuous supply through physical loss control. This allows optimizing for service improvement.

9.3.1 The interaction between physical losses and hours of supply

Understanding the link between physical losses and hours of supply starts with understanding why water is "rationed" through intermittent supply. It is often assumed that such rationing reduces consumption – and indeed, it can initially (Charalambous, 2014). However, studies show that a move from rationing to continuous supply is as likely to reduce consumption as to increase it, at least in the medium term. Another study found an increase in consumption after shifting to 24/7 supply in four Indian cities (Andey & Kelkar, 2007). Yet another study found there is no correlation between continuity of service and consumption (Dahasahasra, 2007). Often, consumers have adapted to rationing by installing distributed storage, such as roof tanks. These solutions allow customers to have water 24 h/day on their premises, even when supply from the network is intermittent. In these situations, creating 24/7 supply does not increase the availability or use of water.

If water rationing through intermittent supply does not reduce consumption, how does it keep the physical use of bulk water in line with the physical availability of bulk water? The answer is that – for any given infrastructure condition – intermittent supply reduces physical losses. To put it plainly: the main reason for chronic intermittency in supply is to reduce leakage.

Table 9.1 provides an example of how this works. Imagine a city with physical production capacity of 640 MLD, which exactly meets the bulk water demand composed of consumption of 400 MLD and leakage of 240 MLD. (This is shown as 'Time 1' in the table).

Table 9.1 Example of interaction between physical losses and hours of supply.

	Time 1	Time 2	Time 3	
	Demand and Supply in Balance	**Demand Exceeds Supply**	**Physical Loss Reduction Option**	**Service Improvement Option**
Hours of supply	24	20	20	24
Bulk water input available (MLD)	640	640	640	640
Consumption (MLD)	400	440	440	440
Physical losses (MLD)	240	200	167	2000
Total of consumption and physical losses (MLD)	640	640	567	640

Rising income levels mean that consumption grows to 440 MLD (shown in Time 2). The network condition remains constant, so leakage would still be 240 ML (given 24-h supply) As a result, total demand for bulk water is now 680 MLD. This exceeds available capacity by 40 MLD.

The utility responds by introducing rotational supply cuts. Initially, consumption may fall, but then returns to previous levels as customers install distributed storage. However, the reduction in hours of supply reduces the physical volume of leakage, so now the consumption demand of 440 MLD can be met with the same input capacity of 640 MLD. Bulk supply and demand are back in balance.

To achieve this balance, the utility will have to reduce average hours of supply to 20 h. This follows from the fact that the system leaks 240 MLD in 24 h, so leakage every hour will be 10 MLD. Assuming for simplicity constant pressure in the network. Reducing the period of supply by 4 h will therefore reduce total leakage by 40 MLD. In this way, intermittent supply can reduce total leakage to 200 MLD, restoring the balance between the supply and demand for bulk water input.

Not satisfied with this situation, the city utility brings in a Performance Based NRW contractor (Time 3). Assume that the contractor reduces physical losses by 33.3 MLD, from 200 MLD to 167 MLD. What happens then? The utility could choose to keep average hours of supply at 20 per day, and so benefit from lower pumping and treatment costs, and deferred investment in future supply (Physical Loss Reduction Option). Under this option, there is now excess bulk supply of 33 MLD.

Alternatively, the utility could increase hours of supply. In fact, in this case, the utility could restore 24-h supply. Following leakage control, the network now leaks 8.33 MLD each hour (167 MLD divided by 20 h). If the network is pressurized 24 h/day, total leakage will be 200MLD (8.33 MLD/h times 24 h is 200 MLD). Consumption is still 440 MLD, so total bulk water requirement is again 640 MLD, exactly equal to the capacity.

In situations of intermittent supply, there is a trade-off between hours of supply and volume of physical losses, as illustrated in Figure 9.5. A physical loss reduction program has the effect of reducing the overall leakiness of the network, and so allowing lower levels of leakage for any given duration of supply – or alternatively, a longer duration of supply for any given level volume of leakage. It follows that the benefits of any given leakage reduction program can be taken either as increased hours of supply, or as reduced bulk water input requirements.

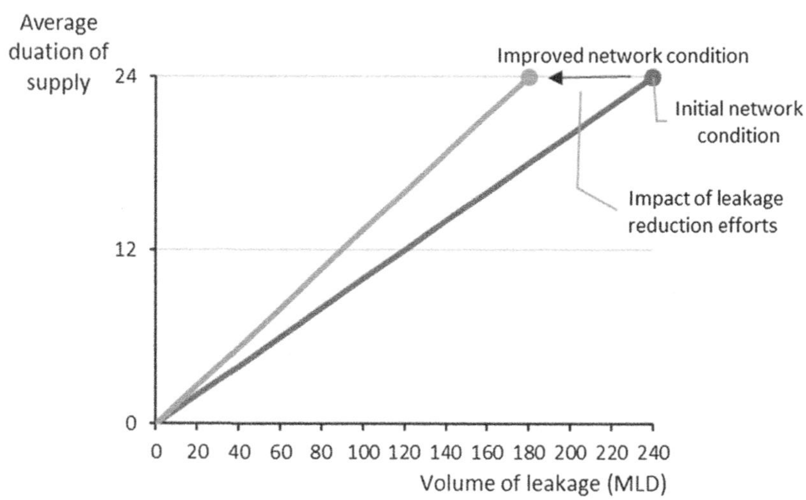

Figure 9.5 Trade-off between leakage and hours of supply.

9.3.2 Valuing benefits from increased duration of supply and increased access

To optimize NRW control in conditions of intermittent supply, we need to put a value on increased duration of supply. We need an appropriate physical measure of duration, and a value of that unit.

There are at least two physical measure of supply duration:

- Average hours of supply across the network
- Number of customers on 24-hour supply (with pressure at or above a specified level).

The "number of customers" metric may be preferable, for several reasons:

- There may be a step change in benefits when 24-hour supply is reached. This is because moving from intermittent to continuous supply can prevent contamination of the piped water – there is thus a gain in quality, not only in convenience. Continuous supply also puts an end to damage to pipes caused by the recurrent pressurization–depressurization cycle of intermittent supply. Instituting continuous supply can also free up staff previously engaged in implementing rotational cuts by opening and closing valves.
- Many NRW experts recommend bringing up areas of the network to 24-hour supply zone-by-zone, so the "customers on 24-hour supply" measure may correspond better with how progress will be made on the ground.
- It is possible that monitoring the number of customers on 24-hour supply may be easier than measuring average hours of supply across the network.

In Figure 9.6 below, optimization of a physical loss reduction program for the number of households receiving 24/7 supply is shown. The value of bringing one more household to 24/7 supply is shown as a horizontal line. The cost of bringing a household onto 24/7 supply is shown as a supply curve that slopes upward as more households are brought onto 24/7. The optimal number of households to be brought onto 24/7 supply through physical loss control is shown as the intersection of these two curves.

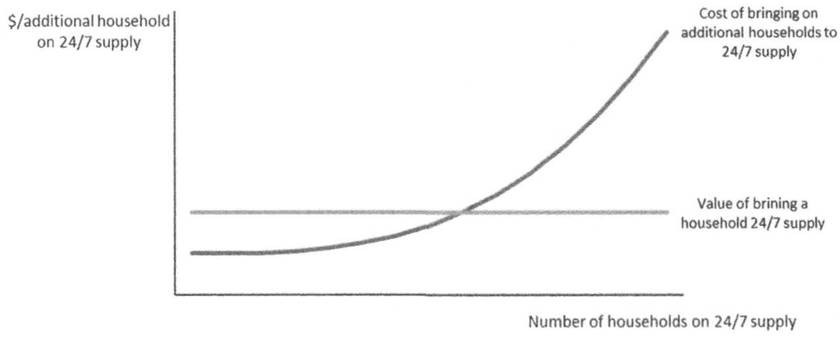

Figure 9.6 Optimizing physical loss reduction for service improvement.

9.3.2.1 The cost of getting households onto 24/7 supply through physical loss control

The curves showing the cost of bringing additional households onto 24/7 supply is derived from the reduction in leakiness of the network necessary to allow a part of the network to be pressurized for 24 h, while holding bulk water input constant. The low hanging fruit of leakage reduction will save a certain quantity of water across the entire network quite cheaply. This water can in principle be diverted to a single zone, in which 24/7 supply can be achieved. In this way, a small number of households can be brought up to 24/7 supply at low cost. However, to save the water needed to bring subsequent zones up to 24/7 supply, the cost increases. Progressively more expensive leakage control measures need to be deployed across the network as more and more zones are brought on to 24/7 supply. This is why the cost of bringing households onto 24/7 supply increases as the number of households so served increases.

9.3.2.2 The benefits of getting households onto 24/7 supply through physical loss control

The value of achieving 24/7 supply, meanwhile, remains roughly constant as more and more households experience improved supply. The benefits of bringing a zone from intermittent supply (for example, water for 12 h every day) to continuous supply (water for 24 h every day) comprise the following elements:

- Greater convenience (or the possibility of not having to install or maintain distributed storage)
- Greater quality (as continuous pressurization prevents contamination of piped supply)
- The value to the utility of reduced damage to its infrastructure from the pressurization–depressurization cycle.

Under the assumption that 24/7 mains supply is optimal, the benefit of bringing a household onto 24/7 supply is the lower of (a) the value to the household of 24/7 supply and (b) the cost of bringing a household onto 24/7 supply through expansion of bulk water supply. It is possible that in some systems that have become fully adapted to intermittent supply – for example systems in some Indian cities – switching to continuous supply from the mains is not cost–benefit justified. This would be because households have already sunk the cost of distributed storage and treatment units, which means they get most of the benefits of 24/7 supply, even though the utility supply is actually intermittent. Whether this is the case should be studied for each system. For the purposes of this discussion paper, we make the conventional assumption that it is cost–benefit justified to transition the entire city to 24/7 supply.

The cost of bringing a household onto 24/7 supply by increasing distribution input can be calculated from the LMRC of water. The amount of water needed for each household will be the amount of additional leakage that occurs in a zone as it moves from intermittent to 24-h supply, divided by the number of households in the zone. As an example, if the city illustrated in Table 3.1 had 2 million households, then the decrease in physical losses per household required to get to 24/7 would be 16.5 L/day, equating to a total saving of 33 MLD. Multiplying this by the LRMC of water calculated as described in Section 9.2.4, and discounting to give a present value, would give the benefit of getting a household onto continuous water supply by means of physical loss control, expressed in dollars per household.

If there is enough water to bring all households onto 24/7 supply, the next step will be to expand the network to serve more households. However, we do not go into optimization for that case here.

9.4 OPTIMIZING COMMERCIAL LOSSES

As a first approximation, commercial loss reduction generates not a reduction in bulk water input required, but an increase in revenue earned for any given volume of water put into supply. For example, if an illegal connection is regularized, the water that was previously taken illegally will now be consumed legally, and charged for. Similarly, if meters that are under-recording are replaced with accurate meters, more of the water consumed will be charged for.

As a first approximation, the value of commercial loss reduction is the increase in revenue it brings. This increase is measured in dollars, not cubic meters. Given certain simplifying assumptions, it is optimal to continue with commercial loss reduction until it costs a dollar in commercial loss reduction activities to generate an extra dollar in revenue, as Figure 9.7 shows.

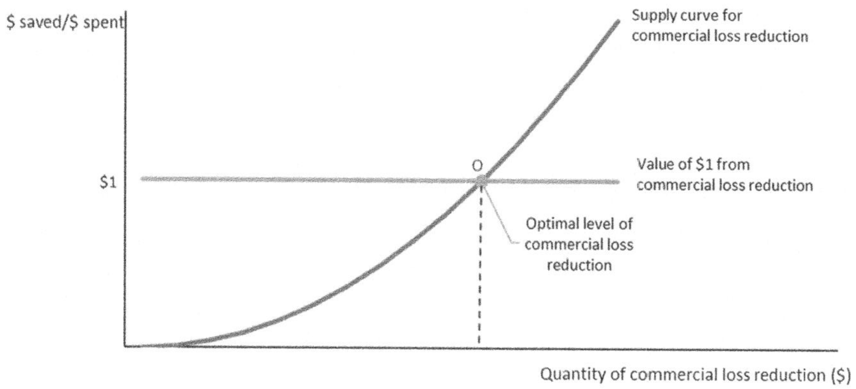

Figure 9.7 First approximation optimization of commercial loss reduction.
Note: As with Figure 9.1, the costs curve represents the present value of the total costs of an intervention, including both initial capital costs and ongoing costs.

Three simplifying assumptions were made in Figure 9.7:

1. There is no change in consumption as a result of the commercial loss reduction program
2. Each dollar billed is collected
3. We are looking at the optimization from the point of the utility, not of society as a whole.

The implications of relaxing each of these assumptions is now explored.

Reducing commercial losses means making people pay for water they were previously getting for free (for example, in the case of regularizing an illegal connection) or making people pay more per unit consumed (for example, replacing under-registering meters). To limit how much more they have to pay, people will cut back on the amount of water they consume. Thus, the benefits of commercial loss reduction are a mix of additional dollars of revenue, and reduced need for bulk water.

The same may be the case when customers are not metered. If customers are charged on the basis of estimated bills, and the bills are under-estimates, the under-estimated amount will be a commercial loss. In this case installing meters will result in some combination of increased revenue and reduced consumption. Of course, if the estimated readings are too high, revenue will reduce for the utility once meters are installed. Meter installation may still be justified on the grounds of demand side managementbecause once metered, customers may reduce their consumption, since doing so can (further) reduce their bill.

Moreover, the effects of collection inefficiency need to be considered. A water utility's revenue is not all collected as cash. Some bills are never paid. So, a dollar of revenue is actually worth just the part of the bill which is expected to be collected.

The viewpoint from which benefits are assessed also needs to be considered. To the extent that commercial loss reduction results in people paying for water where previously they consumed but did not pay, the utility benefits, but arguably society as a whole does not. From a social point of view, benefits come when the water is consumed. The value that someone gets from the water, does not change when a bill is issued for something that was previously unbilled – it is just a transfer. However, in this analysis we look at optimization from the perspective of the utility, for which a dollar in extra payment is a dollar of benefit.

In summary, reducing commercial losses benefits the utility both by increasing revenue and by reducing consumption. The reduction in consumption is a benefit to the utility because the water consumed has an value equal to the long run marginal cost of bulk water production (as established in Section 9.2). The increase in revenue should be measured by additional dollars collected. These two monetary values can be added together to give a total value of commercial loss reduction. The optimal level of commercial loss reduction then occurs when an additional dollar spent on reducing commercial losses yields exactly one dollar of combined benefits from increased cash inflows and reduced bulk input water demand.

9.5 A GENERAL OPTIMIZATION FRAMEWORK

Table 9.2 summarizes the benefits that NRW control programs can bring in terms of reduced bulk water needs, reduced energy and chemical costs, increased duration and quality of service and additional cash received. It also shows the cost of the NRW reduction program, in capital, operations and maintenance, and program management terms.

Table 9.2 Benefits and Costs of NRW control program.

	Units	Value per Unit	Subject to
Benefits of NRW Control			
Reduce bulk input required	MLD or m^3	LRMC of bulk input capacity or volume	Constant volume, pressure and hours of supply
Increase duration of supply	Households on 24/7 supply at or above a specified pressure	\$ per household (derived from cost of 24/7 supply using increased bulk supply)	Constant bulk water input
Increased quality of water consumed	Households receiving water that is safe to drink	\$ per household (derived from health studies)	Constant bulk water input
Increase in cash received	\$	\$1	Constant bulk water input
Reduction in energy used in pumping and treatment	kWh	\$/kWh at applicable tariff	Constant volume, pressure and hours of supply
Costs of NRW control			
Capital costs of NRW control	\$	\$1	
Operating and maintenance costs of NRW control	\$	\$1	
Program management costs of NRW control	\$	\$1	

Taking all these costs and benefits into account, it is possible to write a general formula for the benefits of an NRW reduction or performance improvement contract.

$$\begin{aligned}\text{Net Benefits} &= \Delta\,(\text{Bulk Input Required cu.m}) \times \text{LRMC}\left(\frac{\$}{\text{m}^3}\right) + \Delta\left(\text{households on } \frac{24}{7}\text{ supply}\right) \\ &\quad \times \text{Benefit per household }(\$) + \Delta\,(\text{cash received}) - \Delta\,(\text{pumping and treatment costs}) \\ &\quad - \text{Capital costs of NRW reduction program }(\$) \\ &\quad - \text{Operating and Maintenance Costs of NRW Reduction Program }(\$) \\ &\quad - \text{Program Management Costs of NRW Reduction Program }(\$)\end{aligned}$$

This equation is subject to certain constraints and assumptions. For example, no value is put on an increase in hours of supply until it reaches 24/7. This might not be right – and certainly there is a cost to reducing hours of supply for any customer below what they were previously receiving. This formulation of the general optimization framework will only work if accompanied by a contractual stipulation that hours of service may not be reduced for any customer.

9.6 CONCLUSIONS

Performance Based NRW Reduction Contracts must specify targets or goals, and payments for achieving those goals. The Optimization Framework provides a way to set goals and payments that will generate maximum value for the public.

Concerning goals, the framework encourages policy makers to think beyond reducing NRW, to the goal that NRW reduction serves. Table 9.3 summarizes the goals commonly served by NRW reduction, how progress toward those goals can be measured (indicators) and the value of such progress in terms of the economic benefits of a one-unit movement toward the goal. The table also describes the situations in which any particular goal tends to be relevant.

Table 9.3 Goals, indicators and values for NRW reduction contracts.

Goal	Indicator (Units)	Value of an Increase in the Indicator	Where Relevant
Reducing total costs of providing continuous supply to everyone in the service area	Reduction in input volume needed to provide universal continuous service (m^3)	Long Run Marginal Cost of Water ($\$/m^3$)	Where universal continuous supply is the norm i.e. most developed countries, but only a minority of developing countries
Increasing the number of customers with 24/7 service, for a given volume of input water	Customers with 24/7 service (number)	$ per customer, derived from the total cost of bringing a customer onto 24/7 through increase in input volumes	Many developing cities, in which service is currently intermittent for some or all customers, but continuous service is a desired and achievable goal
Increasing the number of customers with intermittent service, for a given volume of input water	Customers with service of at least some defined level of frequency (number)	$ per customer, derived from the total cost of water required to add a customer at the defined services level by expanding bulk supply	Cities in which some combination of extremely scarce water resources and unusually low income makes intermittent service the economically optimal solution
Increase in cash collected, for a given number of customers and quantity of consumption per customer	Additional cash collected ($)	An additional $ collected would be valued at 1$ by the utility	Systems where the utility lacks funds to sustain and expand service, and there are material gains in commercial efficiency possible

The framework shows that the optimal level on any of the goals is the level at which the cost of adding an additional unit exceeds the value of the unit added. This framework thus helps policy makers to think about the goals to set for an NRW reduction contract, and the payments that it would be reasonable to make to achieve those goals.

It also provides a way in which performance-based NRW reduction contracts can be linked to expected economic benefits. The formula presented in the general optimization framework can be used both to predict the total economic costs and benefits of a contract, and to track actual economic benefits and costs compared to those predictions.

9.7 NOTE

This paper was first drafted as an input to development of the World Bank's program, "Developing Good Performance Based Contract Practices in the Marketplace to Manage Non-Revenue Water".

9.8 REFERENCES

Andey S. P. and Kelkar P. S. (2007). Performance of water distribution systems during intermittent versus continuous water supply. *Journal of the American Water Works Association*, **99**(8), 99–106.

Charalambous B. (2014). The challenge of dealing with intermittent water supply. IWS TG Workshop Dealing with Complex Interrelation of Intermittent Supply and Water Losses, Portugal.

Dahasahasra S. V. (2007). A Model for Transforming an Intermittent into a 24 × 7 Water Supply System. www.geospatialtoday.com (last accessed 1 September 2016).

Dickinson M. A. (2007). An Approach to Determining the True Value of Lost Water: California's Avoided Cost Model. Alliance for Water Efficiency. IWA Water Loss 2007 Conference Proceedings, Vol. 1, 23–26 September. See http://www.amac.md/Biblioteca/data/25/17/01/03.2.pdf

Tripartite Group (2002). Best Practice Principles in the Economic Level of Leakage Calculation.

Chapter 10

How to improve water services performance

Brice Cabibel, Fabien Mainguy and Florie Lozivit

Public–private partnership (PPP) is about transferring risk to a private operator, hence all PPP contracts are fundamentally performance based, even though the recent terminology tends to restrict the terms "performance-based contracts" (PBCs) to contracts in which only some operational tasks are passed to a private operator paid partly on results and the utility remains under public management. However, the transfer of risks and associated performance-based framework needs to be thought through carefully if a government wants to attract private operators on financially sustainable contracts. Based on its worldwide and diversified experience, SUEZ has broadly identified "10 commandments" for sustainable PBCs. A strong water and sanitation governance: (i) definition of clear objectives for the PPP; (ii) explicit distribution of roles between the public authority, the utility and the contractor; (iii) adequate devolution of responsibilities to empower each actor on its role and objectives, and ensure that it fulfill them. A smart PBC design: (iv) focus on a limited number of KPIs; (v) provide a reliable baseline or the possibility of a re-baselining; (vi) define ambitious but SMART targets and indicators (specific, measurable, assignable, realistic, time-related); (vii) ensure due mitigation of risks, with risk management responsibility given to each actor who can best handle it; (viii) allow flexibility in the management of CAPEX and for an integrated management of CAPEX and OPEX. A resilient economic model: (ix) enhance financial sustainability of the PPP by complementing tariffs if necessary with taxes and transfers; (x) promote a sustainable local anchoring with the involvement of the water service stakeholders, and a focus on knowledge transfer (both technical and managerial skills) to local staff.

10.1 INTRODUCTION

Water demand will continue to increase in developing countries in the coming years due to their rapid urbanization and the evolution of their living standards. To match this demand, new water infrastructure (plants and networks) may well be needed. But firstly, upgrading the water service performance will be

Authors: Jan Janssens, Didier Carron, Philippe Marin, Philip Giantris and Tom Williams
doi: 10.2166/9781780405964_0089

above all necessary to optimize existing assets, so as to avoid or delay additional capital investments. Hence, there should be a shift in strategy, from an infrastructure-based approach, to a more services and results-oriented approach.

As a world leader in operating water and sanitation services, including in water-stressed areas, SUEZ is strongly focused on service performance. *Performance-based contracting* (*PBC*) is a concept which aims at preserving the water resources by optimizing assets at three levels: plants, networks and customer management. PBCs encompass all types of contractual models (Operation and Maintenance – O&M, Management Contract, Alliance, Integrated Services Contract, Lease Contracts, Concessions, etc.), provided that adequate *Key performance indicators* (*KPIs*) are set out and jointly followed up by the public authority and its contractor.

The critical success factors for sustainable PBCs are:

- **A strong water and sanitation governance** which clearly defines stakeholders' roles and responsibilities and the framework for collaborating to deliver essential services.
- **A smart PBC design** which includes a baseline and fair performance incentives, leaves room for contractual flexibility, and empowers the operator on CAPEX/OPEX management.
- **A resilient economic model** with enhanced financial sustainability, risks mitigation instruments, and a sustainable territorial anchorage of the water utility.

There is no silver bullet: well-designed PBCs are heavily customized.

10.2 A STRONG WATER AND SANITATION GOVERNANCE

Water challenges are generally linked to governance issues. Improving sustainable water and sanitation services for all, involves partnering with multiple players. The establishment of clear objectives, an explicit distribution of responsibilities and a predictable financial framework are necessary conditions for all stakeholders to work together and reach the said objectives. Water governance means implementing a framework for effective and efficient water policies, fostering stakeholder's engagement and a cross-sectoral vision (OECD, 2015).

The public authority is responsible for supplying citizens with equitable access to safe and affordable water and sanitation services, defining a workable institutional environment for integrated water resources management: regulation, strategies, financial resources (tariffs, taxes, and transfers) and appropriate contractual and supervision modes with the operator. Under balanced and rightly regulated public–private partnership (PPP) arrangements, the public authority remains the decision maker on all matters including tariffs and taxes, and the owner of the water production and distribution assets.

Its capacity to efficiently manage sector regulation, as well as contract monitoring and evaluation, is a key element to guarantee sustainable quality and value for money of water and sanitation services.

The operator acts on behalf of the public authority as a professional expert to run the services. It should not supersede the authority and must act in a transparent manner. The operator shall bring advanced expertise and know-how resulting in improved operational performance of the utility and optimization of the overall cost of service.

The private operator can either support the utility, in the case of service contracts, or manage it. In both cases, having a smooth interface between the PBC contractor (the operator), the utility's staff and the public authority is paramount to ensure collaboration and performance.

This collaboration must be carefully organized, with particular attention to be given to:

- *Utility ownership and mutual commitment*: the utility staff must understand the benefits for them and be willing to cooperate with the operator.

- *Ring fencing*: the interface between the operator and the utility operations/activities needs to be carefully defined.
- *Sustainable financial policy* (tariffs, taxes, transfers).
- *Capacity building and knowledge transfer*: the utility staff must fully understand the contractual and technical approach.

Transparency, integrity and the engagement of stakeholders, especially civil society and NGOs, contribute to ensure shared benefits for all parties and to create a climate of trust and accountability.

BOX 10.1 THE ALLIANCE MODEL, AN INNOVATIVE GOVERNANCE, AUSTRALIA

In 2011, the South Australia Water Corporation (SA Water), a state-owned organization which serves 1.5 million people, and the Allwater Joint Venture (Transfield Services, SUEZ) signed a 10-year Operations and Maintenance *Alliance contract*.

The Alliance model is based on *co-governance and profit sharing*, secured during the procurement process:

- The Alliance Management board (AMB) is driven by a co-operative "Alliance Leadership Team" (ALT) made up of representatives of the public authority and of the operator. Decisions are preferably consensual. To guarantee acceptability and cultural alignment, *the Alliance Management Team (AMT) is composed of staff from both parties*, who work together on a daily basis to manage and improve the water system of the Alliance.
- This co-governance model ensures both parties are aligned on objectives, scope, targets, indicators and action plans. The project base costs (direct and specific overheads) are covered and constitute the minimum revenues for the operator. Coverage of off-site overheads and possible profit depend on a *variable fee, linked to the achievement of KPIs*: the gain/pain share is based upon 400 Specific Performance Indicators (non-financial performance), of which about 30% can trigger penalties/incentives. KPIs are related to people & safety, product quality, customer service, environmental performance, asset maintenance, business enhancement, etc.

The Alliance model establishes co-construction, joint responsibility, enhanced trust, transparency, and no blame/no dispute relationships. Both parties are encouraged to work co-operatively to overcome operating issues on a cost-effective basis.

10.3 A SMART PBC DESIGN

A Performance-based contract creates strong *incentives for efficiency in conditioning the contractor's remuneration on achieved results.*

Most PBCs focus on water network efficiency and the "non-revenue water" (NRW) indicator, which encompasses commercial and physical losses. The rate of NRW is generally used to assess the general performance of a water utility: reducing NRW requires an integrated approach of asset management, active leakage detection, operational performance, customer management and financial management.

However, beyond this specific focus, *PBCs can be applied on a holistic level* and a wide variety of activities in the water utility's value chain (treatment plants, networks, billing and collection), for improving either operational efficiency (reduce costs), service quality, capital works or environmental compliance.

More generally, the concept of performance can apply to all kinds of contracts, *once the private operator has been in a management/operation position* for several years and provided that the contract strikes the right balance between OPEX and CAPEX.

Figure 10.1 shows the structure of PBCs:

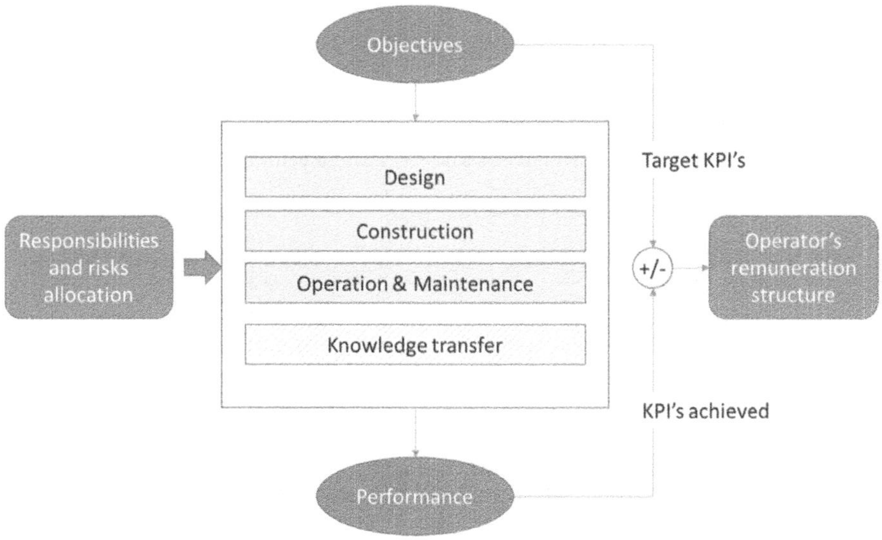

Figure 10.1 Performance-based contract diagram.

10.3.1 Baselines, targets and incentives

Before committing to binding targets and taking the risk of financial penalties, the PBC contractor needs to analyze its risks starting from a *reliable baseline* consisting of a detailed, sincere and objective dataset and vision of what the initial situation is in terms of the financial, operational and technical performance of the utility.

(a) Where the utility has developed a "data room", an information system, with reliable and strategic information on the water and sanitation networks and equipment, the baseline would consist of the data room.

(b) When reliable data is available but not computed into a data room, the project preparation process shall include a *Detailed Project Report* (DPR), generally prepared by independent consulting organizations. This report shall provide sincere operational figures (commercial, technical, economic, etc.) on current operational levels, and set realistic targets as well as an estimation of the CAPEX which would be necessary to meet these targets. The bidding documents shall include a Bill of Quantities (BoQ) for the capital works.

(c) When data is not available or not reliable, a *competitive dialogue* is recommended to allow potential bidders to conduct their own due diligence in order to appraise the current situation and discuss with the client on contractual targets.

Given that the parties may have different views and interests, they need to agree on objectives and scope, broken down into targets and indicators, to assess water services performance progression throughout the contract period. *Targets and indicators should be ambitious but SMART* (Specific, Measurable, Assignable, Realistic, Time-related) and freely accepted as such by the contractor in reference to the baseline and market conditions.

The day-to-day operation and performance of water services are monitored and evaluated through KPIs reported on a regular basis (as per the conditions of the contract). It is preferable *to limit the number of KPIs leading to bonuses or penalties to facilitate the monitoring process.*

PBCs should incentivize operational performance by introducing *a balanced profit sharing mechanism*: the variable remuneration based on performance should not exceed 20 or 30% of the principal contract amount.

BOX 10.2 PERFORMANCE-BASED MANAGEMENT CONTRACT, JEDDAH, SAUDI ARABIA

In 2008, the National Water Company (NWC) contracted Jeddah Water Services (JWS), a Joint-Venture between SUEZ and Aqua Power Development to manage the Jeddah City Business Unit (JCBU) through a 7year management contract serving a population of more than 3.5 million inhabitants.

This contract was part of a national policy to develop PPPs in the water and wastewater services, to increase their performance in a context of extreme water scarcity (98% of the resource comes from desalination). It was strongly performance-oriented and comprised the following features and objectives:

- A world class service in Jeddah: a 24/7 access to drinking water; a division by three of emergency response time to repair leaks (from 60 h to less than 24), a fight against wastewater overflows within the 1,000 km wastewater network.
- Transfer of knowledge to the local staff of 1,400 employees.
- 20 Key Performance Indicators (KPIs), covering the whole scope of the contract from service continuity to leak run times and availability of wastewater pumps, which lead to performance bonuses or penalties. Internal reporting and external benchmarking measures are reported for information, but do not lead to any additional or reduced fee.

The contract execution has been successful and it was estimated that around 25 million m^3 of water had been saved after three years of operation. The leak run time was reduced from 60 h to 16 h. To achieve these results, SUEZ implemented its Operation Control Center which provides efficient network surveillance. A training centre has been set up and a thorough training program proposed to the utility's staff.

No.	KPI Name
1	Customer satisfaction index
2	Complaint resolution time (Breakage & Leakage)
3	Quality of water (Testing compliance)
4	Average time to install water connection
5	Average time to install wastewater connection
6	Non Revenue Water
7	Volume of water saved due to leak detection
8	Water supply
9	Billed water quantity
10	Number of new WW connections per year
11	Total Revenue
12	Environment Violation Revenue
13	Number of environmental violations
14	TSE Revenue
15	Total cash collection (current year + previous year)
16	Average Revenue Yield
17	Opex/Billed m^3 (Excluding depreciation)
18	Lost Time Incident Frequency (Per 1 Million Man-hour)
19	Training Days/Employee
20	Employee Satisfaction Index

10.3.2 Ensure mitigation of risks and leave room for flexibility

A thorough risk assessment must be carried out at the beginning of a project so that a balanced and appropriate allocation of risks is performed between the various stakeholders, with the purpose of identifying tools that can be activated efficiently and promptly should the project materially diverge from initial projections (Figure 10.2). The discrepancy from initial business plans can result from various factors including:

- The economic context within the country or linked to a more global context, e.g. drop of oil prices resulting in a sharp cut in public expenditure and subsidies to basic services operators;
- The political, regulatory and institutional contexts which can significantly change over the life of long-term assets (20 years or more) such as water treatment and distribution facilities (e.g. upgrade in potable water quality requirements);
- The operational aspects in case of unexpected challenges: pollution around water intake, social turmoil, climate events, pollution norms and controls for wastewater management, bankruptcy or quality defects of subcontractors, etc.

Figure 10.2 Illustration of allocation of risks in project finance (source SUEZ/OECD, 2015).

Lack or insufficient mitigation measures might deter operators from bidding on projects that require long-term commitments within unsteady environments. SUEZ looks forward to building a balanced risk allocation and mitigation system, as well as sufficient contract flexibility that ensures room for maneuver in future negotiations.

In particular, the contract should set the framework of future renegotiations *to adjust to unpredictable conditions or inaccurate information.*

Ideally, the PBC should include a first phase of diagnosis (be it part of a competitive dialogue or after contractor selection) where the operator would: (i) confirm the baseline; (ii) define the action plan; (iii) implement methodologies and complex systems such as customers information systems, SCADA, etc. and (iv) launch the engineering and procurement steps for works.

If the actual situation is significantly different from the contractual baseline, a *re-baselining* should be allowed, with subsequent renegotiation of the capital works program, KPIs, targets and incentives. An *independent consultant/third party* can be contracted during this phase to objectify the discussions, give technical support to the client and facilitate the decision-making process.

Flexibility also means that the operator, which is mandated to reach the targets, is empowered regarding its organization and action plan, *particularly in the implementation of the capital works program*, where it should be fully accountable on results and responsible for the selection of subcontractors (KPIs focused on results and not on methodology), within a general framework preventing excessive overruns.

BOX 10.3 CONTRACTUAL INNOVATIONS AND RISK MANAGEMENT LIMITATIONS, KARNATAKA, INDIA

In 2015, the state of Karnataka issued three tenders for implementing sustainable 24/7 water supply in the cities of Hubli-Dharwad, Belgaum, and Gulbarga (respectively third, fifth and sixth largest cities of Karnataka in terms of population), as part of the Karnataka Urban Water Sector Improvement Project supported by the World Bank. The Hubli-Dharwad contract takes the form of an *O&M for the complete water cycle under a 12-year contract*. Capital investment is financed by the state, the municipality and the World Bank

Some of the contractual innovations provided are:

- The contract is split into three phases (baselining 1 year, transition period 3 years, sustaining period 8 years). *With a 1-year study period* provided for the operator to carry out various surveys, design, capital works estimation and an O&M plan, bidders are not misled by uncertain baseline data on assets and customers in the tender submission.
- The operator is given freedom to *optimally use the Capex Envelope*, in terms of utilization across assets in order to achieve KPIs.

However, in this case, SUEZ (and other major operators) decided not to bid for this project due to the risk allocation which was deemed to be unbalanced:

- The liability cap of the entire contract value (i.e. 12 times annual revenue) would have represented an unacceptable burden to SUEZ. In this case, a letter of support from the parent company could have been sufficient with the liability cap limited either to 10% of the contract value or 1 year of revenue.
- The 3-year period to complete all works was deemed unrealistic given the market conditions (weak capacity of civil works subcontractors) and the difficulty to undertake road works in such densely populated cities. In this case a duration of 5 years was deemed preferable.
- The public procurement procedure imposed to select civil works subcontractors would have entailed additional delays, lack of flexibility and possible quality issues. In this case, where the operator is accountable on results, full control on subcontractor selection and monitoring should have been granted.
- KPIs were far too optimistic in terms of the NRW target of 15% with a 24/7 service operation after three years. Failing to reach them would have led to substantial financial losses for the operator as the performance-based incentive was 40% of the contract amount. SUEZ's recommendation is that this variable fee be no more than 20% of the contract amount, and that KPIs be realistically achievable (A of SMART).

10.3.3 Develop an integrated approach

An integrated approach in the utility is more likely to enhance its overall performance. This approach is implemented through comprehensive & multi-faceted PBCs which impact all the utility's departments (strategy, commercial, production, operations, procurement, human resources, finances) deploying a diverse expertise (management, technical & financial).

Similarly, *an integrated approach of the assets' lifecycle* is to be promoted in PBCs. Adequate level of OPEX has to be projected over a long-term period matching the life of the assets. *The Operator shall manage both OPEX and CAPEX* in order to be fully liable for the performance level achieved.

The *duration of the contract must be sufficient* to realize, control and optimize necessary works and operations. PBC projects must be sustained over the long-term (≥10 years) to have durable impacts and infuse throughout the utility.

BOX 10.4 INTEGRATED CAPEX/OPEX MANAGEMENT, SEAAL, ALGIERS

In the 2000's, Algeria was facing a critical situation regarding access to drinking water and sanitation. Water scarcity and obsolete infrastructure prevented the utility from providing reliable access to services. Facing severe droughts that exacerbated social pressure, the Algerian State launched large-scale institutional reform and called upon PPPs.

The public company in charge of water and wastewater services across Algiers, the "Société des Eaux et de l'Assainissement d'Alger" (SEAAL), was created in 2006 with the objectives of modernizing water supply, sanitation and customer services. SEAAL immediately transferred to SUEZ, under a 5-year Management Contract (renewed in 2011 and 2016), the operational management of the utility.

After only 4 years, the 3.8 million people living in Algiers were supplied with water on a continuous basis (against 8% in 2006 (Figs 10.3 and 10.4)) and by 2015 the sanitation coverage had increased from 6% to 60%. Between 2006 and 2015 the number of customer connections increased from 422,000 to 730,000 and the tariff collection rate from 70% to 93%, resulting in a surge in turnover with a flat tariff, and making the PBC self-financing. SEAAL is now seen locally as one of the most efficient public services of the city.

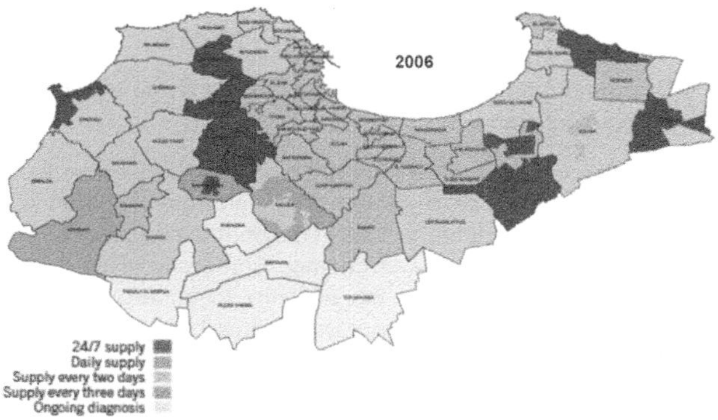

Figure 10.3 Wilaya of Algiers: water distribution situation in May 2006 (65% with 24/7 supply).

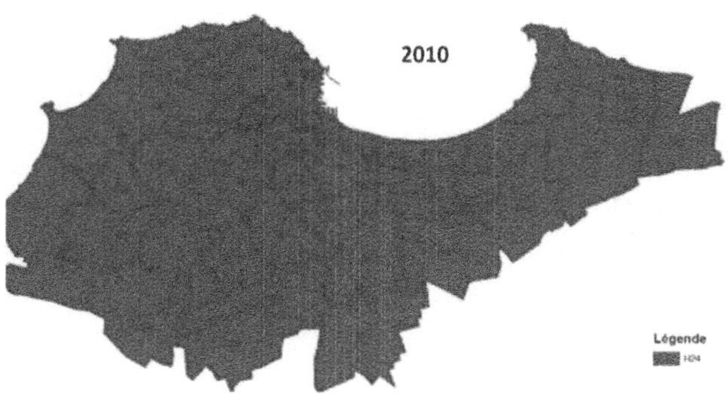

Figure 10.4 Wilaya of Algiers: water distribution situation in April 2010 (100% cover).

These results have been possible thanks to an integrated management of CAPEX and OPEX serving a long-term vision for continuous improvement.

- *SUEZ had sufficient autonomy and available funding in CAPEX and OPEX to optimize the management* of the water supply system, even though the main structural investments before or during the contract (dams, water supply treatment and desalination plants) were financed and/or carried out by the Government. SUEZ had the liberty to make its choices regarding maintenance and asset renewals. The selection of subcontractors was made by SUEZ and *not bound to public procurement procedures*.
- The management contract yielded substantial *savings on operating expenses* (OPEX) compared with the business plan, owing to better productivity and cost-efficiency.

Long-term commitment and collaboration of the different stakeholders (based on the action plan and the business plan) has been essential to achieve substantial and sustainable results.

10.4 A RESILIENT ECONOMIC MODEL

10.4.1 Enhanced financial sustainability

In developing countries, the pressure on financial resources (national and local authorities), generally coupled with low tariffs, often results in an under-financed water and sanitation sector, to the detriment of service quality. By nature, the financial sustainability of water utilities is hardly ever granted when initiating a PBC, yet it is of utmost importance to ensure sustained results and stimulate private sector investments in water and sanitation utilities.

Effective financial planning requires finding the relevant mix of revenues (tariffs, taxes, transfers, the OECD's "3 Ts)" to allow for sustainable cost recovery – at least operation and maintenance costs, if not capital investments amortization. The mechanisms for covering these costs and their evolutions over time must be clearly set out between the public and private stakeholders so as to maintain an affordable pricing mechanism tailored to local constraints. Diversifying resources enables innovative experiences for a more balanced distribution of local income.

BOX 10.5 PROPERTY DEVELOPER'S TAXATION: LYDEC'S EXPERIENCE, CASABLANCA

Addressing urban development challenges in Casablanca, exacerbated by climate change and aging networks, require financing means that are out of reach of local capabilities. To address this issue, the city of Casablanca created a taxation on real-estate developments to finance their connection to water and sanitation services. This resource is also used for general investments such as structural infrastructure, network extension, land acquisition and social connections. The tax is collected by the City of Casablanca and deposited in a "Works Fund account", which is managed by SUEZ's subsidiary, Lydec, on behalf of the public authority.

Given the evolution of urban needs, the level of contribution of real-estate developers has been revised three times between 2009 and 2014. The part of investments financed by the Work Fund contributions' revenue has risen from 17% in 2004 to 56% in 2014 (Figure 10.5).

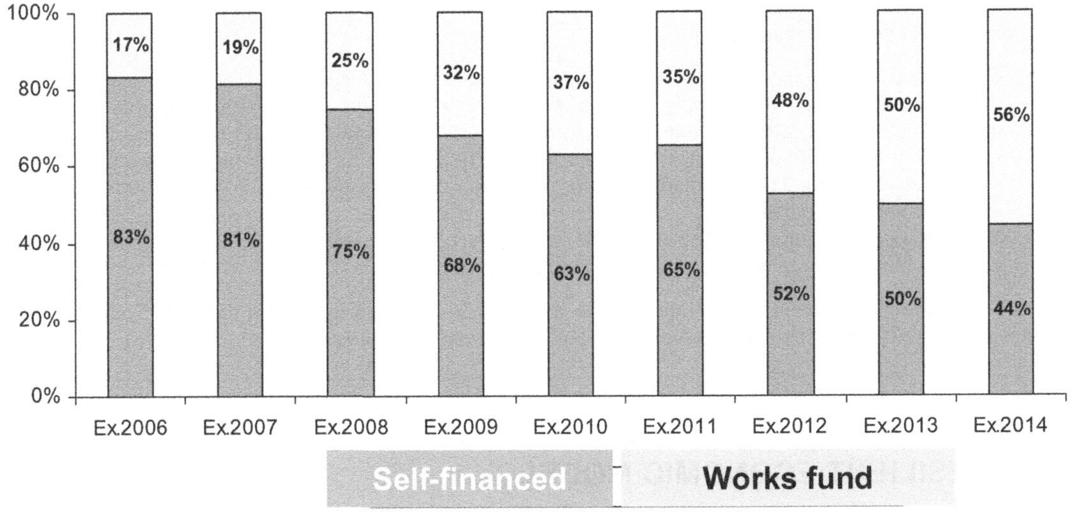

Figure 10.5 Lydec's investment budget, origin of funds.

In parallel, the tax is pretty well accepted by the developers as it represents only 1% of the sale price of houses, which remains very limited. The major advantage of this original financing source is its flexibility, its transparency and its fitness-for-purpose as it is perfectly synchronized with needs created by the rapid urban development.

10.4.2 A sustainable territorial anchorage

Water services are key to improving the living and health conditions of people, protecting the environment and contributing to a territory's development. *As responsible corporate citizens,* water utilities interact with multiple stakeholders, directly or indirectly impacted by the services, to prioritize strategies according to their needs. They should pay particular attention to vulnerable local communities (women and children, disabled people, disadvantaged people, etc.).

BOX 10.6 INNOVATIVE MODEL TO DEVELOP ACCESS TO WATER IN LOW-INCOME AREAS: PALYJA, JAKARTA

In charge of clean water production and distribution in Western Jakarta, SUEZ's subsidiary PALYJA has been extending water services to low-income areas since 1998.

Giving the high level of groundwater contamination, with seawater infiltration and non-treated wastewater pollution, thousands of non-connected inhabitants from deprived neighborhoods had no other alternative to satisfy their basic needs but to buy clean water from ambulatory vendors, at a price more than twenty times higher than PALYJA's tariff.

In order to supply these areas with safe and affordable water, PALYJA implemented the specialized "Water for All" program of SUEZ. Based on a combination of social engineering methods and water-related expertise, this program led to the creation of three specific ways of improving water accessibility in poor areas:

- *Individual domestic connections* with alternative funding, like the *Global Partnership on Output Based Aid (GPOBA)*, whose subsidy was linked to two output indicators: number of individual connections realized; and average water consumption per month higher than 10.8 m^3 during three consecutive months, to make sure that connections were actually used and reached their intended purpose. This project allowed PALYJA to connect more than 5,000 households.
- Implementation of *master meter systems* to provide clean water either in legal settlements or illegal settlements. Benefitting from an important involvement of local communities and NGO's, this scheme benefitted 576 households in north Jakarta.
- **Water kiosks with tanks supplied by trucks.** This scheme aims to improve the access to clean and affordable water where the network is not yet extended. Trained managers are in charge of the distribution of water, at a transparent and affordable price, by using pushcarts. A total of 52 water kiosks were installed, supplying more than 7,000 households on a daily basis.

Beyond the transfer of knowledge and expertise, which is a priority, PBCs aim at a longer term strengthening of skills and at *changing the utility's operational and managerial culture.* A successful transition involves consolidating operational and managerial practices, as well as developing leadership and ethics. A 10-year contract is considered a minimum to achieve such successes. Capacity building should also include *local service providers and contractors* (supervision, specific training), who would be incentivized to invest in the local market. In this regard, a good practice is to seek a balanced *international/ local partnership*, giving the right balance of expertise and competitiveness.

BOX 10.7 STRENGTHENING SUSTAINABLE OPERATIONAL AND MANAGERIAL SKILLS: SEAAL, ALGIERS

To ensure long-term sustainability of the Management Contract results, SUEZ was particularly challenged by SEAAL on knowledge transfer.

SEAAL benefited from the extensive expertise of SUEZ with the on-site presence of senior directors and professional experts (around 30 expatriates) and 800 days/year of technical assistance on strategic topics.
SUEZ used two corporate knowledge transfer methodologies aiming at strengthening both operational and managerial skills.

- Based on a professional practice framework, *WIKTI (Water International Knowledge Transfer Initiative,* Figure 10.6*)* assesses water services performance and proposes mid-term improvement plans to be discussed with the authorities, while consolidating and harmonizing local occupational skills on all daily activities through an extensive training plan (more than 137,000 days in Algiers since 2006; 18,000 days/ year).
- The *OPT (Optimizing Personal Talents)* tool supports the empowerment of the public utility through managerial performance. OPT is based on a managerial charter inspired by international standards and adapted to local socio-cultural profiles. By defining individual roadmaps for all top managers (250 people), this methodology aims at spreading common managerial references and good practices within the utility, which contribute to transforming the managerial culture.

Local teams also have access to specific training tools (e-learning, mentoring, etc.), integrate with SUEZ's international and diversified network, benefit from SUEZ R&D and innovations, etc.

Such a knowledge transfer approach ensures an efficient and modern management, maintains collective mobilization and a shared common culture. It represents an attractive and complete support mechanism for a public structure seeking to rapidly improve its effectiveness and to capitalize on the latest industry knowledge.

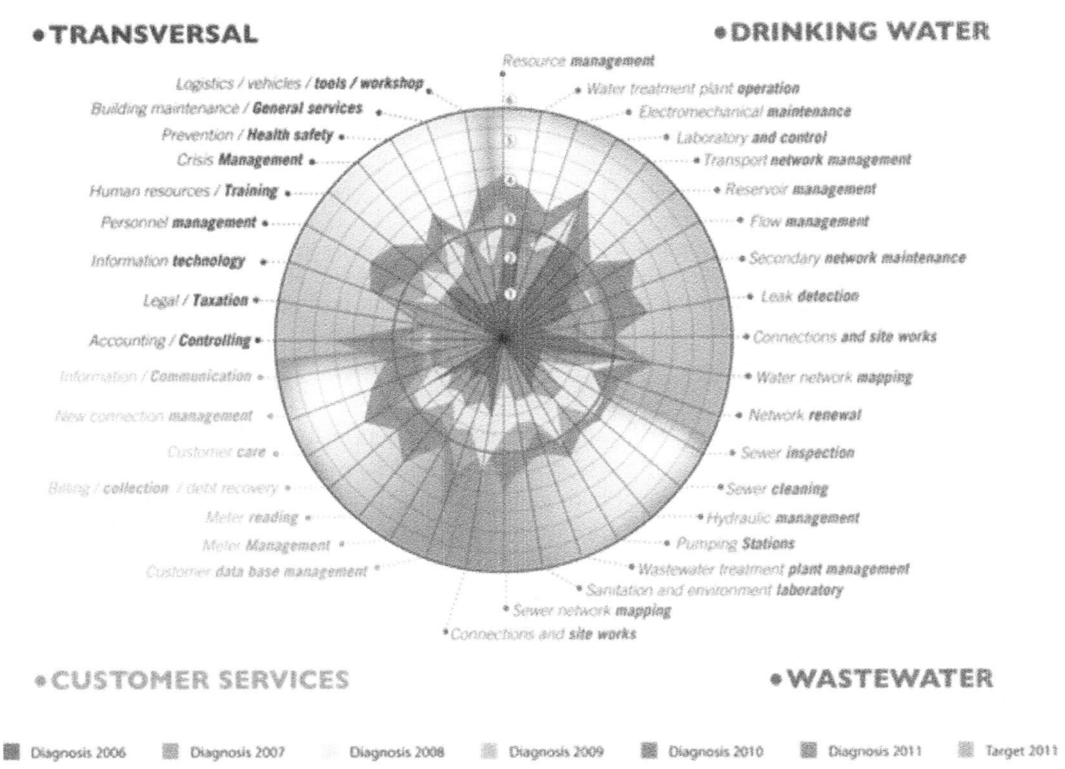

Figure 10.6 WIKTI/The 38 principal daily activities.

Lastly, raising *citizen's awareness* about the best practices in water consumption and conservation is also a key role of the utility, which increases its fame and legitimacy among individual customers.

10.5 CONCLUSION

The adoption in 2015 of the 2030 UN Sustainable Development Goals, with one out of 17 goals dedicated to water and sanitation, is putting water challenges at the top of the global political agenda. To face the challenge of "ensuring availability and sustainable management of water and sanitation for all" (Goal 6, UN, 2015), public authorities must be proactive, ambitious and able to assume their responsibilities, starting from building institutional capacity and allocating relevant financial resources.

The intervention of a private operator through a performance-based contract (PBC) can be helpful to bring international expertise and fast track water services improvement.

Building a win–win PBC involves a close collaboration and trust-based relationship between the public authority, the water utility and the operator.

The operator needs the targets and key performance indicators (KPIs) to be jointly determined based on a reliable baseline which could be built in a first contractual phase or during a competitive dialogue if not already existing. Enabling the operator to implement an integrated vision, facilitated by contractual flexibility and an adequate risk mitigation framework, leverages the benefits of the public–private partnership (PPP).

Sustaining PBCs' results requires: (i) enhancing the financial sustainability of water services by ensuring a long-lasting and rightly sized mix of funding sources (tariffs, taxes, transfers); (ii) building local skills through an ambitious knowledge transfer program; (iii) and work on social acceptability through continuous dialogue with the water service stakeholders.

10.6 REFERENCES

OECD (2015). OECD Principles on Water Governance. OECD, Paris. See http://www.oecd.org/governance/oecd-principles-on-water-governance.htm (accessed 24 November 2017).

UN (2015). Sustainable Development Goal 6. See https://sustainabledevelopment.un.org/sdg6 (accessed 20 November 2017).

Chapter 11

Performance-based contracts (PBC) and regulatory issues

Sanford V. Berg and Robert N. Gakubia

One mechanism employed by regulators to improve the performance of commercialized utilities is the use of performance-based contracting. This involves the setting of monitorable targets to be achieved within a strict time frame by the utilities with incentives for achievement and sanctions for non-achievements. The performance targets include, among other things, the minimum service levels, management and operational efficiency, and utility improvements. The performance targets should be agreed by all the parties concerned and be part of the country's vision on the progressive realization of the right to water services. This chapter addresses performance-based contracts (PBCs) and regulation in the Kenyan context. It discusses the tools and key performance indicators employed by the regulator, the challenges as well as incentives and sanctions applied to enhance the performance of utilities.

11.1 BACKGROUND

In Kenya, The National Water Policy of 1999 and the Water Act 2002 (Act) triggered extensive reforms to Kenya's water sector, bringing it in line with international best practice. The main objectives of these reforms were to improve water resources management, meet growing demand for water services, attract more professionals into the sector, attract greater investment, and create a modernized sector that was more robust and more capable of responding to the emerging challenges such as climate change and urbanization. The key reform features included: separation of policy from other functions; separation of water resources management and water services provision; separation of regulatory functions from investments and operations; separation of asset holding from operations; increased user participation; enhanced pro poor orientation; socially responsible commercialization in the provision of water supply and sanitation services. Conflict resolution was conferred by the act to the Water Appeals Board as an alternative to legal procedures.

Authors: Jan Janssens, Didier Carron, Philippe Marin, Philip Giantris and Tom Williams
doi: 10.2166/9781780405964_0103

In 2010, Kenya promulgated a new constitution. Fundamental to the new constitution was the creation of two levels of government; the national government and devolved governments (county governments). The ownership, use and regulation of water resources, consumer protection and national public works was assigned to the national government while county governments were assigned water service provision, sanitation, catchment management and county public works.

Further the constitution entrenched the right to water and sanitation in the bill of rights, effectively making water and sanitation a human right. These developments created the need to align the Water Act 2002 to the constitution. Consequently, a new act – the Water Act 2016 come into effect in April 2017 and is under implementation. In the Kenyan context the attainment of the progressive realization of the right to water depends on three crucial elements, namely, investment level, performance of the utilities and orientation on demand (targeting), seen in terms of service improvement to the poor within a wider context of effective governance. The allocation of resources and ensuring this is linked to the sector investment needs is anchored at the policy level. The other two aspects of ensuring increased focus to the poor and the continued improvement in performance by the utilities is within the domain of the utilities and the county governments. The role of regulation in this arrangement is to ensure the progressive realization of this right within a framework that protects consumers and the environment and helps to reconcile the social and economic interests. It also involves performance monitoring of utility operations according to set standards, enforcing the licenses issued to them and reporting their performance to the public, all this so as to ensure compliance and progressive improvement in service provision.

11.2 MONITORING AND EVALUATING UTILITIES

A key part of a performance-based contract is the determination and evaluation of whether the parties are fulfilling their obligation. To this end, Wasreb (Water Services Regulatory Board) has developed regulatory tools that include: a performance assessment system, an information collection system, targets for sector benchmarks, and service provision agreements (licences) that incorporate a time table for achievement of minimum service levels (MSLs) specific to a utility. Other tools include tariff guidelines, corporate governance, consumer engagement, business planning and reporting guidelines.

Performance assessment is carried out on the basis of nine KPIs. These are Water Coverage, Drinking Water Quality, Hours of Supply, O + M Cost Coverage, Personnel Expenditure as a percentage of O + M Costs, Revenue Collection Efficiency, Non-Revenue Water, Staff Productivity and Metering Ratio.

The choice of the nine KPIs was based on the main goals of the sector: quality of service, economic efficiency and operational sustainability. Figure 11.1 shows the weights allocated to the clusters.

11.2.1 Building information systems with key performance indicators (KPIs) that enable regulators to monitor, and evaluate operators

The Water Regulation Information System (WARIS) is the regulator's main tool for data collection. The system is an online database with one central shared database. The system also has a provision for offline data entry, which enables the importing and exporting of the data to and from the central database with internet availability. The system has been designed to ensure that data protection and integrity is maintained. In order to improve on the quality of the data that is submitted, the system provides for validation checks – both internal and external; and provides an audit trail for tracking of any changes made.

Once the regulator makes a data request, the utilities activate the specific datasets and provide the data required within an identified time. Figure 11.2 presents the data collection cycle within WARIS.

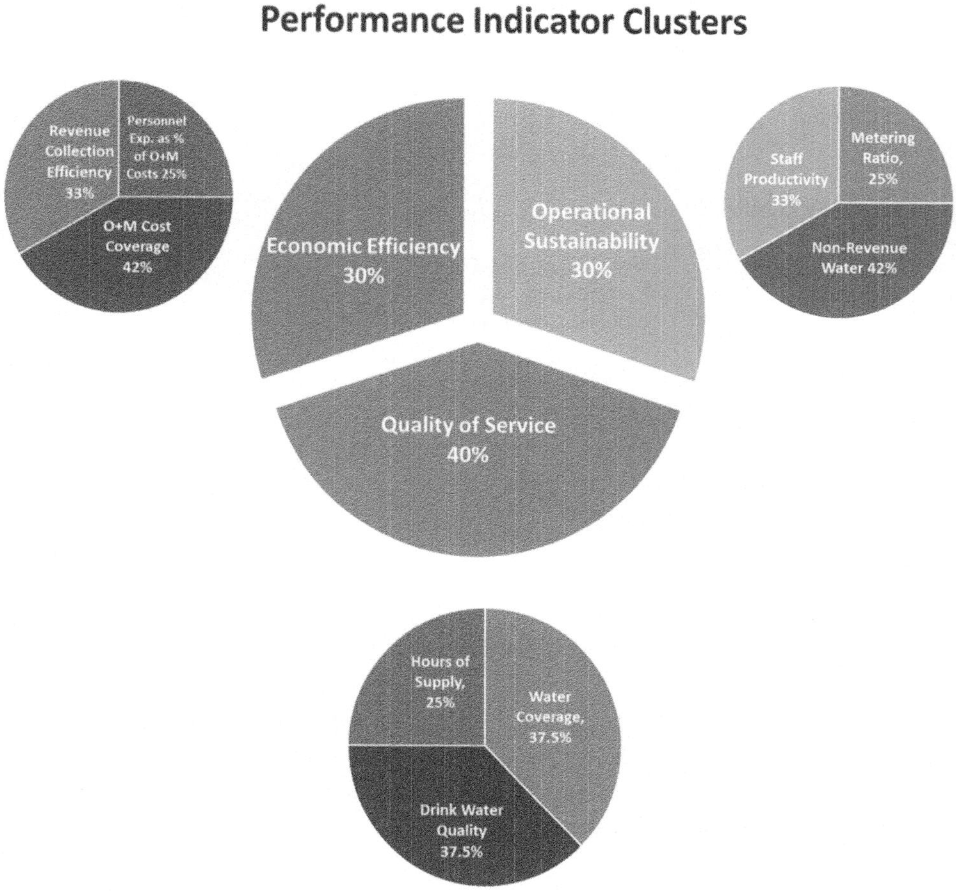

Figure 11.1 Weights for the indicator clusters.

BOX 11.1 BENEFITS OF WARIS

- Increased transparency in the sector
- Facilitated publishing of periodic sector performance review reports
- Improved accessibility of data and information for monitoring, planning and decision making
- Support for effective management audits
- Support for the monitoring of the performance of utilities, Water Service Boards (WSBs) and county governments in investments and water service provision

WARIS has the ability to provide information on 67 indicators both for water and wastewater in the different areas of utility operations, namely, technical, financial, commercial and human resources. The system also has a provision for reporting on pro-poor performance which covers the identified low-income areas. In order to improve the accuracy, the system is linked to the national census data, from which the annual population projection rates are also obtained, alongside relevant data. This enables the determination

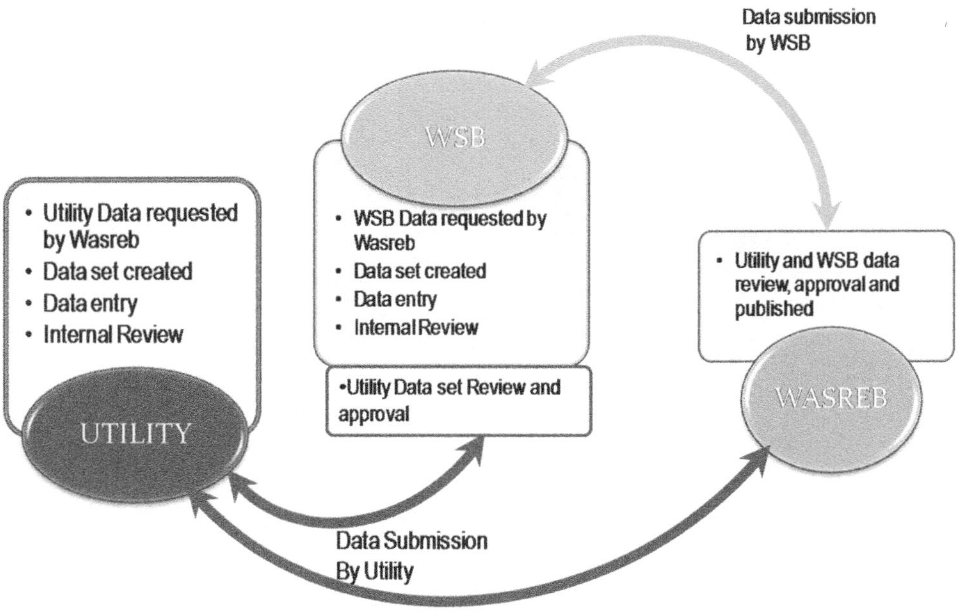

Figure 11.2 The data collection cycle within WARIS.

of population in the service areas of the utilities to be determined with precision. Reports pertaining to the indicators can be generated at all levels. The reporting feature is enriched to highlight improvement, stagnation or declining performance for the different indicators. National performance assessment is based on the nine key performance indicators which are used to rank comparative performance as well.

11.2.2 Setting performance targets and monitoring performance with limited data

The use of a robust data collection system is crucial in this process. Despite the use of WARIS, data flow from the utilities is not a seamless process. Among the key factors that impact on data submission by the utilities are:

(i) Inadequate capacities within the utilities to make use of the reporting tool
(ii) Lack of baseline data for some utilities
(iii) Inconsistent data submission depending on the purpose.

To improve on the quality of data submitted, Wasreb corroborates the data that is submitted through WARIS with data from inspections and the tariff application process.

11.2.3 Developing incentives for quality improvements, cost containment and network/service expansion

Considering that the provision of water services is largely monopolistic in nature, comparative performance reporting is a means to spur competition. To improve on the competition however, it is crucial that good performance is incentivized while poor performance is sanctioned. The role of incentives in this case being to promote, encourage, motivate or obligate the actors to exercise their mandates with clear objectives. Wasreb has developed sector benchmarks for the KPIs that considers three stages of performance, namely, good, acceptable and not acceptable (Figure 11.3). The limits for each performance level have been defined based on international good practice while considering the current status of the Kenyan water sector. The sector benchmarks are dynamic and reviewed on the basis of performance improvement of the sector.

KPI CLUSTER		Indicators		Sector Benchmarks: Good	Sector Benchmarks: Acceptable	Sector Benchmarks: Not Acceptable	Scoring Regime: Performance	Scoring Regime: Score
Quality of Service	1	Water Coverage, %		>90%	80-90%	<80%	≥90%	30
							≤50%	0
	2	Drinking Water Quality, %		>95%	90-95%	<90%	≥95%	30
							≤90%	0
	3	Hours of Supply, No.	Population >100,000	21-24	16-20	<16	≥20	20
							≤10	0
			Population <100,000	17-24	12-16	<12	≥16	20
							≤6	0
Economic Efficiency	4	Personnel Expenditure as Percentage of O+M Costs, %	Large and Very Large Companies	<20%	20-30%	>30%	≤25	15
							≥35	0
			Medium Companies	<30%	30-40%	>40%	≤30	15
							≥40	0
			Small Companies	<40%	40-45%	>45%	≤40	15
							≥45	0
	5	O+M Cost Coverage, %		≥150%	100-149%	≤99%	≥150%	25
							≤90%	0
	6	Revenue Collection Efficiency, %		>95%	95-85%	<85%	≥95%	20
							≤85%	0
Operational Sustainability	7	Non-Revenue Water, %		<20%	20-25%	>25%	≤20%	25
							≥40%	0
	8	Staff Productivity (Staff per 1000 Connections), No.	Large & Very Large Companies	<5	5-8	>8	≤5	20
							≥8	0
			Medium & Small (less than 3 towns)	<7	7-11	>11	≤7	20
							≥11	0
			Medium & Small (3 or more towns)	<9	9-14	>14	≤9	20
							≥14	0
	9	Metering Ratio, %		100%	95-99%	<95%	100%	15
							≤80%	0
Total Maximum Score								**200**

Figure 11.3 Sector benchmarks and scoring regime.

The sector benchmarks are used for performance assessment of utilities and hence are the basis for determination of incentives and sanctions. The use of incentives and sanctions drives utilities to perform beyond their commitments in minimum service levels (MSLs), which is part of their condition for licensing. Incentives regulation by Wasreb takes the form of recognizing the performance of the utility in its annual *IMPACT* report, granting the utility an enhanced tariff incorporating minor investments and allowing it to leverage on the tariff to access blended financing. Sanctions would take the form of naming the utility in the *IMPACT* report, penalties, fines, denying the utility a tariff increase to putting it under a special regulatory regime (close monitoring).

Generally, quality improvement on services sets in with the attainment of minimum service levels because that is where utilities are likely to slacken their effort, reasoning that they have reached their apex due to the monopolistic tendencies for lack of competition. Regulation raises standards and efficiency in operations, hence utilities and customers get value for money on account of its effective application. The effective application of incentives and sanctions to the utilities impacts positively on service delivery.

Effective PBC application requires concretized empirical data to be the baseline for target setting. Wasreb's consistent publishing of the annual water services sector report ("Impact") has laid the foundation for the application of PBCs, as it establishes baseline data of good quality that allows the setting of authentic targets.

Chapter 12

Peer-to-peer partnerships operational for sustainable water services

Siemen Veenstra

Over the last 10 years, WOPs have been initiated to enhance peer-to-peer capacity development amongst water utilities. Their flexible responsiveness to demand, their strong focus on operational peer-to-peer support, and their focus on financial sustainability are unique selling points. WOPs thereby demonstrate their potential to increase operational performance and create a sound platform for sustainable investments. Critical success factors for WOPs are: an institutional regulatory framework (the enabling environment) that promotes a competitive drive for utilities to perform, a utility management responsive to (local/regional) stakeholders, including customers, and a clear corporate and employee commitment towards utility turnaround: a change in organisation, operations and individual behaviour. As such, WOPs do not require a "grand design", but a demand-driven start, a management commitment to change, and an impact-based approach. Once these are internalised, performance improvement plans may lead to success. Change will take time, however, on a road map of 8–10 years, substantial performance improvements can be harvested and sustained. An example is Lilongwe Water Board of Malawi. After 5 years of service contract, their partnership with VEI resulted in a self-financed performance-based WOP. Drafting of this PBWOP was based on in-depth knowledge of the ground situation, mutual trust and shared confidence that performance can be improved.

12.1 INTRODUCTION

The poor state of water infrastructure is often a result of many years of neglected O&M, under-investment, as well as inefficient management and resulting operational practices. The challenge is huge when facing the unprecedented urbanisation taking place worldwide, particularly in sub-Saharan Africa. For a long time, measures taken by governments to address service coverage gaps have concentrated on building new infrastructure with little attention given to improving efficiency and productivity of water utilities. However, estimates of finance requirements for water and sanitation expansion point to large funding gaps

Authors: Jan Janssens, Didier Carron, Philippe Marin, Philip Giantris and Tom Williams
doi: 10.2166/9781780405964_0109

and prospects of private sector investments appear bleak. These realities have compelled major players in the water sector to seek alternative approaches to improving water service coverage.

A performance decline of water utilities basically starts with low tariffs and low revenue collection rates. This creates financial stress which constrains operation and maintenance and postpones investments. Services deteriorate, cost for O&M tend to rise, while customers are less willing to pay. Utility managers lose autonomy and incentives while government is interfering in their daily business operations. This in the end results in a system where assets go "down the drain" and service levels degrade to beyond acceptable levels (Baietti *et al.* 2006).

Decades of trial and error have resulted in water utilities suffering from high water losses, leading to low credit scores, and lending agencies and institutions unlikely to finance. Losses reduce the amount of water made available to customers and jeopardise the financial position of a utility. In countries with both water scarcity and a shortage of funds, this is a "disguise" as a growing population is in need of potable water. It is not only disappointing for utility managers, it also questions the role of respective governments and the international financing community that they, in spite of long-lasting engagement and massive investments and sector reforms, were not able to transform poorly performing utilities into financially healthy enterprises that deliver reliable and affordable water services to customers. In 2016, this is not really rocket science.

Twinning has long been practised among water utilities, aiming to capacitate utilities and improve operational performance. Sharing operational experience and best practices is often financed through ODA or CSR funds. In spite of its good intentions, all too often, tangible outcomes are disappointing. Twinning assumes willing partners, an institutional framework, and a regulatory framework creating an enabling environment triggering utilities to perform. The basic idea of twinning is that utility capacity can be enhanced by peer-to-peer interaction between professionals in the field. However, facing the sector's mounting challenges, operational twinning alone is largely incapable of addressing the poor performance properly. The search is on for formats to induce sector reform, address deteriorated infrastructure due to lack of maintenance, corporate culture and related working attitude, fast urban growth, and resulting investment needs that until today coexist with poorly run utilities, artificially low tariffs, and scarce fiscal resources (Marin, 2009).

In the 1990s, a series of public–private partnerships (PPPs) were introduced as a more solid approach to enforce sector reform and utility change by forced "business wise operation", while anticipating substantial private investments to rehabilitate and extend water infrastructure. These expectations did not fully materialise (Marin, 2009). Private financing of urban water utilities remained quite limited and relationships between (often public and private) partners frequently deteriorated over time because of insufficient alignment of objectives. Sometimes ending the partnership was the only solution left. Various governments (responding to society and the public at large) grew dissatisfied with the modus operandi of PPPs, and felt they could better solve the problems by regaining direct public control (Marin, 2009). A typical example is Paris, where in 2008 its Municipality ended private sector contracts with Veolia and Suez Environment that had operated Paris water services since 1985. Publicly owned Eau de Paris, which took over from 2010, has since become an inspiring model for a string of French and foreign cities and a threat that civic leaders now use to force down prices in private contracts.

While privatisation turned out not to be a guarantee for success, neither is ineffective twinning with often bureaucratic and state-run water utilities. In a search for new powerful formats, the United Nations in 2008 called upon water operators to engage in water operator partnerships (WOPs) to share and disseminate good practices, contribute to the MDGs (Millennium Development Goals), and raise the creditworthiness of urban water utilities. IWA and UN-GWOPA took a lead in mobilising the huge potential of predominantly (but not exclusively) public member utilities to engage in WOPs (IWA & UN Habitat, 2009). Since then, numerous WOPs have been created. Their experiences are widely analysed and documented in order to share lessons learned and ensure that WOPs became more powerful than previous twinning arrangements (UNGWOPA database).

Simultaneously, international financing institutions regained interest in performance based contracting as the way forward to impose change upon water utilities. In this format, a (service) contractor is recruited and assigned with a contractual obligation to achieve certain utility KPIs for which a performance related compensation PBC (bonus/penalty) can be obtained. According to Vitens Evides International, this format is powerful but underscores the crucial role of the utility itself to take ownership and responsibility for achieving their KPIs (Tremolet, 2015).

This chapter highlights some of the unique selling points of WOPs, and provides insights into how to develop a successful roadmap to irreversibly change the corporate culture and organisational setting of water utilities to encourage "business wise operations" in providing reliable, affordable and sustainable water services. This chapter is largely based upon the experiences obtained by Vitens Evides International, The Netherlands, over their 10 years of engagement in WOPs all over the world. It tries to synthesise the unique features of *public twinning* and *private involvement* into a hybrid way forward: *professional engagement.*

12.2 THE WATER OPERATOR PARTNERSHIP

The basic instrument of any WOP is capacity development. Capacity-building is based on sharing skills, knowledge, experience, and customer focussed behaviour arranged in partnership with renowned water utilities that have proven track records of good performance. This is recognised as a promising way to improve performance of water operators and accelerate their progress towards MDGs (or Sustainable Development Goals (SDGs)). At operational level, WOPs seek to bridge the capacity gaps that exist through intense face-to-face systematic knowledge and experience sharing with experts of mentoring operators (Figure 12.1).

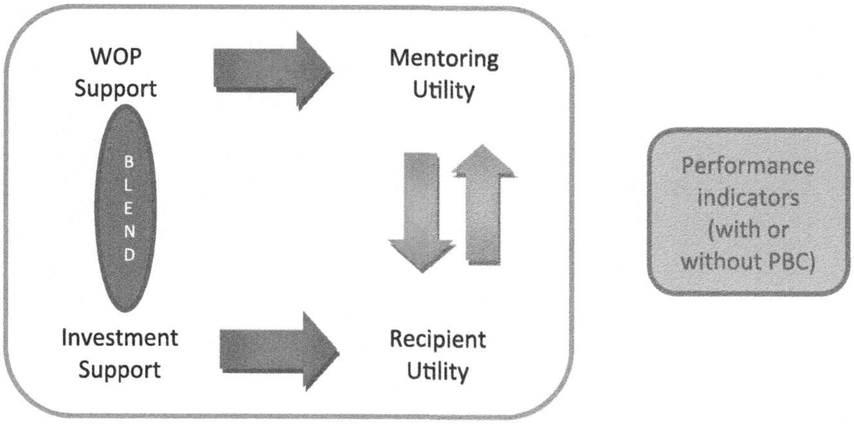

Figure 12.1 WOP format including blended operational and investment support.

The professional engagement between mentor and recipient water utility is normally to document their shared ambitions in a contractual arrangement. There is a gradual conversion from "input driven" to "output based" partnerships. Based on a baseline assessment of utility strength and weakness, performance improvement plans (PIPs) are drafted, framing the WOP activities. The ultimate management mandate for WOP interventions is by default with the recipient utility. Conflict resolution by "marriage counselling" is recommended to mitigate an "early divorce".

Although the use of partnerships is widely studied, its added value it is not often practiced. All too often, the mentor is held solely responsible for the outcomes and impact of WOPs. Beach *et al.* (2005) confirmed that the benefits of partnerships in the UK industry can be quite substantial (Table 12.1), but these are largely (and only) achieved by full alignment of all partnering parties.

Table 12.1 Tangible effects of partnering in UK industries (Beach *et al.* 2005).

Area	Improvement Achieved	Detail
Capital cost	10% reduction	Lower bidding prices and fewer disputes
Construction time	10% reduction	Reduced claims and litigation
Predictability	20% increase	Fewer program overruns
Defects	20% reduction	Improved quality and design
Accidents	20% reduction	
Productivity	10% increase	Lower supervision cost
Turnover & profit	10% increase	Fewer cost overruns

To ensure and sustain improved operational performance a utility turn-around involves decisive leadership and organization of business processes in a way that optimizes the interests of civil society, customers and employees. This is visualized in Figure 12.2 which is based on a model developed at the European Foundation for Quality Management (EFQM).

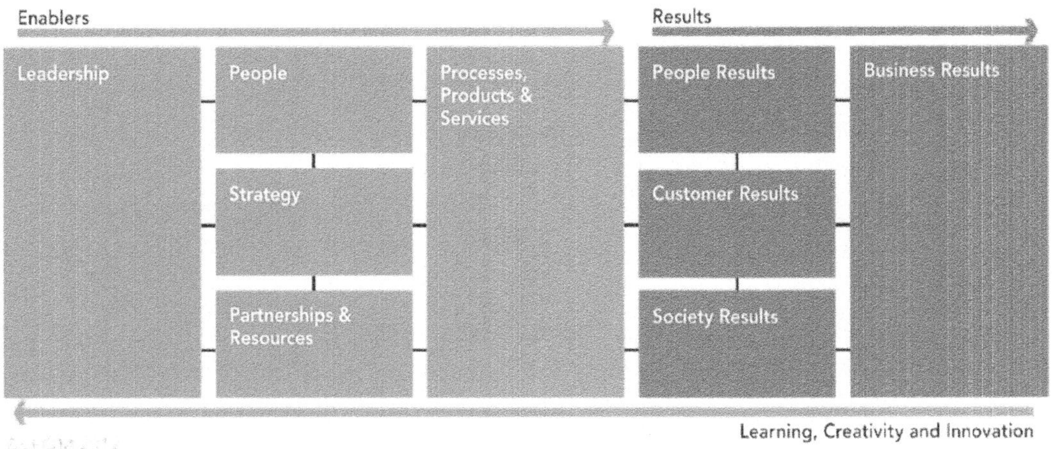

Figure 12.2 A utility turnaround composed of a variety of elements.

12.3 COMPONENTS OF AN OPERATIONAL PARTNERSHIP

The strength of a WOP lies in the simple fact that it is a peer-to-peer interaction at various levels between partner organisations being engaged in the same business. Peer-to-peer interaction within the (field) operations do sort out more impact and transfer of experience, skills, and public behaviour than written plans and strategies.

Basically, what VEI acquired in their more than 15 WOPs in varying institutional settings is that a WOP has four core components of interaction (Figure 12.3).

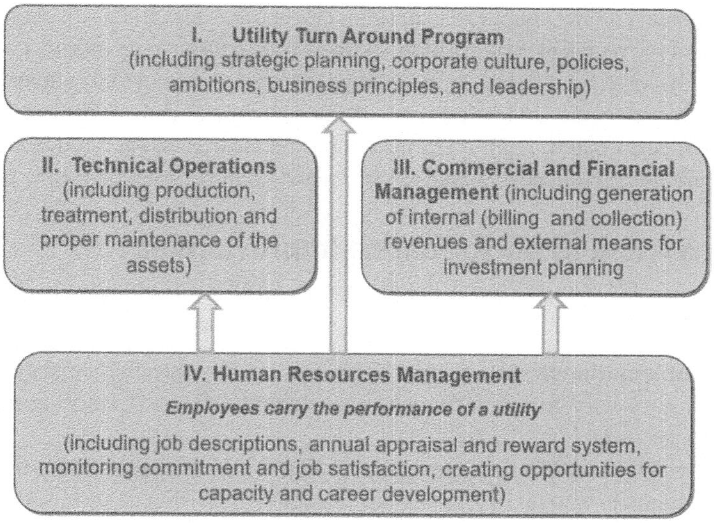

Figure 12.3 Core components of a water operator partnership (after Veenstra, 2015).

While, all too often, attention and corresponding resources gear towards component II, it may turn out to be a huge sink if not properly embedded in I, III and IV. Examples of weak utility management, mis-investments (in e.g. high tech or "cheap" materials), reduced life cycle times of assets due to neglected O&M, or non-committed employees are numerous, and easily explain the current utility performance.

For VEI, any starting point for considering a WOP is on the demanding partner. What are their ambitions and how do they want to realise this. Is there a critical self-reflection and are they ready for a utility turnaround program based on communication and interaction with stakeholders, including employees and customers?

12.3.1 Preparing a WOP

The preparation stage is critical for the success of any WOP partnership. During this stage, their self-reflection and a SWOT analysis is transferred into an actions plan. KPIs are to be identified and aligned to their corporate plan, while contractual specifications with the mentor (as on mandates and conflict resolution) are to be made. In this phase, partners share mutual confidence and trust in corporate strategies and personal integrity.

Before entering into a WOP contract, a thorough review and evaluation of the utility's assets and liabilities, their performance (baseline), as well as of the corporate commitment to change is carried out. No matter how skilled contracting partners are, they can only be efficient and effective in providing tangible assistance if the local utility has fully "internalised" the agreements.

12.3.2 Incentives: yes or no?

To enhance accountability and transparency of WOPs, performance targets may specifically be defined; their achievement over time is monitored to determine if they translate into performance-based compensation (PBC). The question is, who should be incentivised: the mentor or the recipient utility? Where most IFIs tend to incentivise external parties, it was Muhairwe (2009) who introduced internal performance contracting to make public utilities work. It is obvious, in order to achieve and sustain any performance improvements, that transformation of corporate culture, change in management style, and adjustments in operational regimes need to be internalised within the recipient utility. Enforcing their buy-in through an incentive system seems to be a valid option, but it is only powerful if targets are fully aligned between mentor and recipient utility. Independent auditing and an institutional check and balance are needed to ensure that results presented are a true and transparent reflection of performance reality. The need for a concise and strong HR department within a utility is now obvious: they are there to manage human resources and align personal interest to utility ambitions.

12.3.3 Priority setting in performance improvement

Already, the first African WOP workshop in Johannesburg (2007) was attended by over 140 water utility leaders. Their WOP–Africa Action Plan identified the following four themes to be focussed on:

- **Management Information Systems:** assist utilities to establish/strengthen management information systems (MISs) necessary for performance assessments and benchmarking aimed at continuous improvement of services.
- **Services to the Poor:** strengthen pro-poor policies and strategies that define financing and operational mechanisms and tariffs that ensure equitable provision of services to all urban residents and accelerate progress towards the achievement of full coverage (MDGs).
- **Human Resources Development:** prioritise capacity development to foster a vibrant water sector through utility-to-utility exchange of know-how/experience and networking on training opportunities and exchange programs.
- **Infrastructure Development and Asset Management:** establish sound asset management principles based on life cycle costing, taking into consideration the (often) clear separation of operational and ownership roles and responsibilities.

12.3.4 Conditional factors for WOP success

While the strength of WOPs lies in the not-for-profit nature of capacity development through peer-to-peer interaction at operational level within the utility, other factors that affect the overall turnaround of water utilities into effective and efficient organisations are the *enabling environment, water utility management,* and *peer-to-peer interaction* as visualised in Figure 12.4.

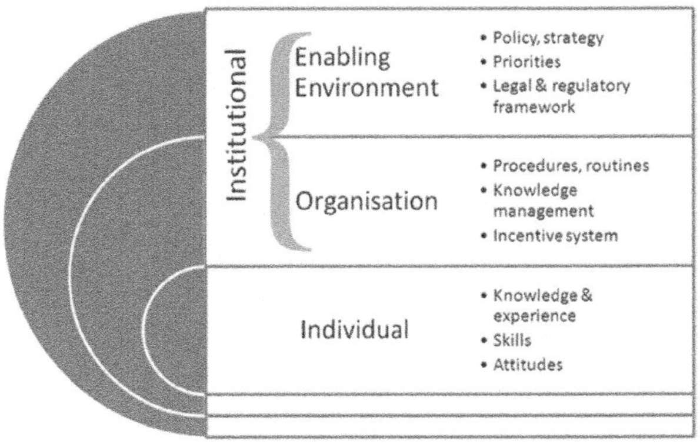

Figure 12.4 Conditional factors affecting the likelihood of a successful water operator partnership Veenstra (2015).

All too often, WOPs solely focus on capacity development, thereby ignoring crucial factors that decide (or at least affect) the efficiency and effectiveness of these peer-to-peer efforts. Utilities face different (and frequently changing) social, political and financial constraints, which affect their operational performance. Institutions, rules and regulations, and political processes determine whether economies grow, whether children go to school, whether development takes off, and whether utilities do perform. Moreover, management is often incompetent to implement appreciative human resources management whereby employees are recognised/rewarded for their efforts by simple appreciation or any other incentives.

12.3.4.1 Enabling environment

A reliable and consistent regulatory framework is intended to create an "enabling environment" for utility managers to manage and improve performance. Governance in general comprises the mechanisms, processes and institutions through which citizens and groups articulate their legal rights, meet their obligations, and mediate their differences (Rouse, 2007). Without effective governance to ensure (a) presence of *institutions* that are able to develop sound policies, (b) clear and *transparent segregation* of policy making, regulation, and service delivery, and (c) access to *financial resources,* very little can be expected from utility operators (Rouse, 2007). One even may argue that poor governance is often worse than no governance at all.

Benchmarking may create some competitive pressure that may trigger utility operators to raise their performance. Not primarily by "public naming and shaming" but by "sharing best practices". It is easily started by self-regulation amongst utilities within a national context. With time, benchmarking may improve and a higher degree of regulation may become desirable to enhance transparency, efficiency, and equity of both privately or publicly managed water utilities. Regulation should aim to delegate management, to maximise the autonomy of utilities and allow them to operate on sound "business principles". Utilities may then shift from crisis management to *strategic asset management* as a sound and solid basis for "running public business".

12.3.4.2 Organisational level

Upon assessing a utility's readiness to embark on a WOP, its strategy and corporate culture are very indicative. If there is a strong and connected management that is based on leadership principles, their commitment to change may be able to overcome resistance. According to Schwartz (2006), utilities that interact most with their customers are the ones that usually perform well. Customer responsiveness and employee satisfaction degrees are also true indicators of the state of the organisation.

Performance-based contracts (PBCs) are negotiated legal agreements between governments, municipalities and water utilities that deliver public water supply and sanitation services. Generally, PBCs are developed to help define the utility development goals. These contracts include time-bound performance targets against which the performance of the operator is measured. If designed properly, performance-based contracts can help lay the basis for the long-term sustainability of the utilities, increasing their efficiency and creating conditions where investment capital can be attracted. Unlike traditional government contracts that focus on inputs (procedures and processes to be used in delivering a service; amount and type of equipment; and/or time and labour to be used), PBCs focus on results, thus encouraging operators to be innovative and to find cost-effective ways of delivering services. However, experiences with PBCs (Hoekstra, 2009) show that no contract, no matter how well designed, can guarantee success during implementation. For this reason, it is important that the parties involved maintain good working relations which can help solve problems in a less formal but sometimes more efficient manner. An independent "marriage counsellor" may be instrumental in managing the contract and partner relations.

12.3.4.3 *Individual level*

Capacity development is often characterised by training and educating employees in technicalities of their work. The focus is on sharing experience and know-how. However, the common weakness in utility performance is in the performance of its individual employees. Often, they are not properly managed, not adequately paid or incentivised, let alone encouraged to "go the extra mile". Thus, it has to do with their behaviour towards work. If not properly guided and embedded in a participatory team effort with management, facilitated by HR departments, employees may choose to resist any change (either on an individual or collective basis induced by, for example, unions). Management is put in a difficult position while the effectiveness of capacity development never translates into better performance (Figure 12.5).

Figure 12.5 Working in a water utility – balancing in a delicate environment.

It is typically in the added value of a WOP that the mentor utility recognises these relevant factors so detrimental to the overall success of any partnership. In Malawi, it took over a year before at peer-to-peer level VEI was recognised as "partner utility" rather than "another consultant passing by". This perceived "peer solidarity" was identified as a turning point in the contract.

The experience of Vitens Evides International further reveals the following ABCD (Alignment of performance indicators, Blending capacity development with investments, Communication with stakeholders, and Design of flexibility) to be essential ingredients of a well-designed and successfully implemented WOP.

12.3.4.4 *Alignment of performance indicators*

Performance indicators allow monitoring of the performance of the operator in a more objective and transparent way. In addition, a bonus and penalty system which is often built into PBCs may be also directly linked to the achievement of these performance indicators. This is one of the reasons why parties to contracts are so concerned with the definition and methodology for setting and measuring indicators, and establishment of a baseline prior to the start.

Indicators should be few and easy to monitor and verify. They should be targeted at the needs of the utility and should reflect the most urgent and critical issues to be solved by the operator. Experience shows that over time, the number of indicators tend to gradually reduce. Health indicators of a utility can be NRW (in absolute terms), WR, and staff productivity (no of staff per 1000 connections). Selection and alignment of WOP KPIs to the corporate KPIs ensure full internalisation and harmonisation of performance targets. This is crucial for the recipient utility and its stakeholders.

Where initial data is often poor (not reliable or available), it is better to set indicators as increments, or improvements defined in terms of percentage above a baseline, rather than as absolute values. Using a percentage of improvement as an indicator makes it easier to integrate modifications to the baseline calculations, when necessary.

Monitoring progress closely and often, with performance indicators makes the system credible and allows parties to the WOP to better understand the challenges as well as encouraging them to seek timely and effective solutions. It may also encourage corporate and/or team spirit within the utility. The powers and responsibilities of any appointed auditor should be carefully defined and balanced with regard to the responsibilities of the operator and the mentor utility. Communication and mediation skills are crucial competencies of any auditor to seriously take the role of "marriage counsellor".

12.3.4.5 Blending with investments

Most water utilities are expected to recover operational costs from the provision of high quality services at affordable tariffs. WOPs often lack resources to provide investment funds to complement their efforts in capacity development. This is a serious weakness of WOPs. International financing institutions, donors and governments will realise that the infrastructure (assets) life time is largely decided by the care taken in operation and maintenance. Therefore, blending investment funds with capacity development will empower WOPs and enhance the financial sustainability of water operators. It was the EU/EIB financed Service Contract in Malawi (2009–2014) that recognised this linkage and conditionally released investment funds according to how the operational partnership with a publicly procured service contractor, VEI, evolved.

In blending the organisation of daily operation and maintenance with long-term investment planning (to keep pace with the fast rate of urbanisation), asset management emerges. Principles of life-cycle costing provides answers on whether or not to replace existing infrastructure. Professional O&M may easily extend asset lifetime by 15 to 35%. This pleas for financial blending of operational and investment funds. A typical *VEI rule of thumb* is that with a 3–5% operations fee the quality and realisation of any investment program increases, the technical and economic life time of the investments rises by double digits, while the financial sustainability of utilities is strengthened. Stakeholders including utility operators are to be challenged to apply sound asset management principles by adopting "business wise thinking" in guiding the way forward to improve water service coverage at affordable cost.

Experience shows that IFIs and donors are more willing to support the water sector when governments introduce accountability and "market" mechanisms, opening up the sector to competition. The release of investment funds may be "conditioned" to the introduction of some form of utility performance benchmarking to increase the effectiveness of transfers.

12.3.4.6 Time will tell

The way forward to improve operational performance and financial sustainability of utilities is, at best, dictated by the way the partnership evolves, rather than by its design format. Flexibility and responsiveness to demand are key features making WOPs a unique proposition. The WOP roadmap is dictated by relationship with management, corporate commitment to change, and alignment of objectives, expectations, and ambitions. Time is, in this respect, a resource, for which the African proverb is: *if you want to go fast, go alone, if you want to go far, go together.* Awarding blended financing opportunities to water utilities in a "cold and distant" procurement process disregards "soft engineering" components of a partnership and may therefore try to arrange "a marriage between professionals", while it ignores the basic ingredient for success: love and affection. To embark on successful and long-lasting cooperation, a gradually developed warm relationship based on mutual trust and respect is crucial to effectively bring about change (Figure 12.6).

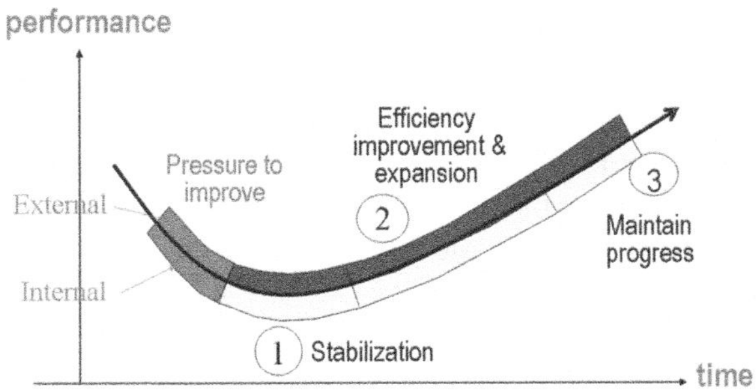

Figure 12.6 Cultivation and sustaining good performance takes time (after Kingdom et al. 2006).

However, improvement by emulation requires that utilities are found that exhibit superior performance and have willingness and specific strengths to match the weaknesses of utilities seeking improvement. A quick assessment can easily provide good indications of who the superior (mentor) performers might be, but more work is needed to confirm their ability to provide peer-support within a WOP context. Governments and IFIs will team up with these utilities to create powerful blending instruments that make them take part in the worldwide challenge: to make the necessary change within those utilities towards more autonomy and financial sustainability.

12.4 SUMMARY AND RECOMMENDATIONS

Their unique selling points make WOPs a powerful instrument that may turn around the financial sustainability of poor performing water utilities. Based on their non-for-profit peer solidarity, capacity development within a utility can be boosted and operational performance be improved. Conditions for success are an enabling institutional environment, a utility commitment for change, and preferably a blending arrangement with investment funds. Key considerations in designing and implementing effective operational partnerships are:

1. WOPs with renowned mentor utilities can be classified as **professional engagement** and may form the best mix of "the good intentions of public twinning" and "the business focus from private involvement".
2. WOPs, provided that they are flexible demand-, and impact driven, can be **powerful to induce and sustain corporate change resulting in operational improvements** within water utilities. WOPs thereby focus on *capacitating managerial and operational staff.*
3. WOPs are based on the joint development of **performance improvement plans** based on a professional and critical baseline assessment and clear expression of the managerial commitment to change. The preparation stage setting the project terms is more guided by establishing "effective collaboration mechanisms" than by negotiating "detailed contract terms".

4. Conditional **success factors of WOPs** are the presence of an enabling environment and a utility commitment towards corporate performance improvement and culture change.
5. **Procurement of WOPs** has to seek a balance between the selection of "best bid for value" and "soft engineering" components. VEI is in that sense sceptical about pursuing contracts that require competitive tendering and rather prefers to negotiate partnership relationships directly with the beneficiary utility.
6. The effectiveness of WOPs can be seriously undermined by the (lack of) leadership and corporate culture of a utility. Utility performance may not improve without an **appreciative working environment** for employees. Appraisals and a subsequent HR response are instrumental to holding employees accountable for their performance and take credit.
7. Whether **performance-based incentive systems** will create an organisational and/or a personal drive to "go the extra mile" is questionable. Without a clear "win–win" for both partners incentive systems tend to fail. Critical issues are the transparent and accountable environment, independent assessment of performance, and uniform translation into compensation (bonus/malus).
8. **Blending of** WOPs with investment funds may accelerate utility performance and improve financial sustainability. WOPs reinforce operational caretaking and preventive maintenance, thereby prolonging the service life time of physical assets. To ensure the recovery of investment loans, a **conditional (performance related) release of investment funds** may support a mentor utility to push for performance improvements.
9. All too often, utilities lack a long-term perspective on tariffs and investment needs. The water sector (governments, utilities and IFIs) therefore needs to create a common and persistent (long-term based) environment to allow utilities "to entrepreneur". This encourages them to **raise financial sustainability**, allowing them to recover (full?) costs and become "bankable".
10. There is growing potential for **scaling-up inter-utility WOPs within Africa**. Contrary to popular perception, the region is not entirely short of well-performing utilities to emulate. These may be cost-effective and well sensitised to prevailing local social and institutional conditions and constraints. Mobilising mentor potential from within regions may definitely have added value.

12.5 REFERENCES

Baietti A., Kingdom W. and Van Ginneken M. (2006). Characteristics of Well-Performing Public Water Utilities. Working Note 9, WSP/PPIAF, May.

Beach R., Webster M. and Campbell K. M. (2005). An evaluation of partnership development in the construction industry. *International Journal of Project Management*, 23(8), 611–621.

Ginneken M. and Kingdom B. (2008). Key Topics in Public Water Utility Reform World Bank Water Working Notes Note 17, August 2008.

Hall D. (2005). Reclaiming Public Water. Transnational Institute & Corporate Europe Observatory.

Hoekstra A. H. (2009). Water Operator Partnerships in Fragile Environments. MSc thesis, Rotterdam School of Management, Erasmus University Rotterdam, The Netherlands.

IWA and UN Habitat (2009). WOPS – Partnerships for utility-to-utility strengthening.

Kingdom B., Liemberger R. and Marin P. (2006). The Challenge of Reducing Non-Revenue Water in Developing Countries. How the Private Sector Can Help: A Look at Performance Based Service Contracting. WSP and PPIAF Discussion Paper Series no 8, December 2006.

Kitano N. and Kenichi A. (2000). Private Sector Participating in Water Supply and Sewerage. Lessons from Ten Case Studies in Developing and Developed Countries. JBIC Review No. 2 November 2000, pp. 68–80.

Marin P. (2009). Public–Private Partnerships for Urban Water a Review of Experiences in Developing Countries. PPIAFF, World Bank, Washington DC.

Metha M. (2003). Meeting the Financing Challenges for Water Supply and Sanitation. WSP World Bank, Washington DC.

Muhairwe W. T. (2009). Making Public Enterprises Work. From Despair to Promise: A Turn Around Account. IWA publishing, London, UK.

OECD (2008). Private Sector Participation in the Water Sector. OECD Checklist for Public Action. OECD, Paris.

OECD (2010). Guidelines for Performance-Based Contracts Between Water Utilities and Municipalities. Lessons learnt from Eastern Europe, Caucasus and Central Asia 14–15 October 2010, Almaty, Kazakhstan.

OECD (2011). Meeting the Challenge of Financing Water and Sanitation. Tools and Approaches. OECD Studies on Water. OECD, Paris.

Rouse M. (2007). Institutional Governance and Regulation of Water Services. The Essential Elements. IWA Publishing, London, UK.

Schwartz K. (2006). Managing Public Water Utilities. PhD thesis, TU Delft, The Netherlands.

Schwartz K. and Coppel G. P. (2011). Water operator partnerships as a model to achieve the Millennium Development Goals for water supply? Lessons from four cities in Mozambique. *Water SA*, **37**(4), 575–583.

Tremolet S. and Mansour G. (2015). Evaluation Report of the EU/EIB financed Service Contract for Peri-Urban Water and Sanitation Services to Lilongwe and Blantyre Water Board, Malawi.

Veenstra S. (2015). Lecture on Water Operator Partnership during course module on: Managing Water Organizations at UNESCO-IHE Institute Delft, 2015.

Veenstra S., Pascual M. S. and Hoffer J. W. (2009). WOPs: It's All About People. IWA Congress, Amsterdam Water Week, Amsterdam, The Netherlands. See www.vitensevidesinternational.com

Veenstra S., Pascual M. S. and Perkins J. (2013). Leadership – Guiding the Way Forward in WOPs. IWA Conference Nairobi, Kenya. See www.vitensevidesinternational.com

Water Operators Partnerships (2009). Africa Utility Performance Assessment. Final Report, June 2009 WSP, UN Habitat, ESAR and AfWA.

Chapter 13

Performance-based contracts in Malawi: Teamwork works

Siemen Veenstra, Alfonso Chikuni and Andrew Thawe

Performance-based contracting (PBC) was the new model to provide technical assistance to poor performing utilities in order to make them do things better in a reasonable period of time. Where management contracts –often private – to operators were commonly received with great (social) resistance and frequently not found very sustainable, PBC was seen as a tough way to trigger an external party to bring about change in operational performance of water utilities without detailed prescription on how to get this done. There was quite some flexibility left to the contracting party to design their "roadmap to the bonus". At the end of the project, results were judged to decide on success or failure, bonus or penalty. EIB embarked on such a format for Lilongwe and Blantyre Water Board in 2008. VEI was awarded the contract after international competitive bidding. Experiences with the PBC model were mixed. Although the flexibility was appreciated by all parties, it was the single focus on KPI results that tended to drift the attention away from the more structural deficits within both utilities: human resources capacity (knowledge, skills and attitude) and corporate culture (management, organisation and leadership). The project format did in the end turn out to be successful in Lilongwe where partners agreed to continue their operational partnership, while in Blantyre the change in leadership did break up the partnership. VEI had mixed feelings at the end as they were forced to focus on achieving results in KPIs rather than on lasting change within the utility.

13.1 THE CONTEXT: HOW THE PBC CAME ABOUT

The Republic of Malawi is with an estimated 16 million inhabitants and a GDP of $224 (World Bank, 2013), one of the least developed countries in Africa. Economic growth has insufficiently been accompanied by transformation of the economy, which still heavily depends on international development funds; around 40% of the government's budget is externally funded (Cammack, 2012). The Government of Malawi intends to provide safe water to their citizens and therefore introduced an appropriate institutional and regulatory framework to enhance financial/operational autonomy of water utilities.

Authors: Jan Janssens, Didier Carron, Philippe Marin, Philip Giantris and Tom Williams
doi: 10.2166/9781780405964_0121

Malawi is amongst the fastest urbanising country in the world, with urban annual growth rates between 4.7% and 6.3% respectively. As a result, over 60% of Malawi's urban population live in unplanned low income areas (LIAs) (Zeleza Manda, 2009). Urban water demand in Blantyre and Lilongwe is rising far beyond current production capacities (Table 13.1). With the observed current climate variations, the need for water rationing is unavoidable.

Table 13.1 Baseline data of Lilongwe and Blantyre Water Boards (2009).

Item	Lilongwe Water Board (LWB)	Blantyre Water Board (BWB)
	Administrative Capital	Commercial Capital
Population served	756,000	875,000
Water production capacity (m^3/d)	95,000	83,000
No of connections		
NRW estimate	39%	50%
Growth rate	3.5%	2.7%
Service level	24/7 gradually turning into seasonal water rationing	Intermittent supplies because of rationing
Staff productivity index	12.8 (461 staff)	14.6 (571 staff)
Main challenges	Inadequate water production capacity to meet rising demand High energy cost (>40% OPEX in BWB) Human capacity weak in operations and management Lack of attention and budget for O & M Lack of accountability on performance (MIS) Low creditworthiness due to high losses No performance culture and staff appraisal Lack of a competitive governance structure (benchmark)	

The Ministry of Irrigation and Water Development (MIWD) is leading WASH policy development and sector oversight. Water Boards, governed by Boards of Directors directly appointed by the Office of the President and Cabinet (OPC), were set up in 1995 as parastatals mandated to provide water and sanitation services, but sanitation is *de facto* still under City Councils (Ministry of Irrigation and Water Development, 2007).

In spite of long lasting involvement, donor cooperation and coordination is poorly aligned. This has created an "ad random environment" for picking financing opportunities "as they come". Private sector involvement was in the late 90s seen as essential to establish autonomy and appropriate incentive structures within Water Boards. The World Bank launched their *Privatization and Utility Reform Project* (PURP) in 2000, introducing a regulator, and opportunities for lease or affermage contracts. The drive for privatising the water sector generated public outcry, while key actors, including water boards, were reluctant to relinquish management control. The privatisation initiative as well as the establishment of a utility regulator was put on hold by the Government of Malawi (GoM). This setting opened opportunities for a less far reaching performance based service contract.

Whereas under a management contract, full managerial responsibility is contracted out to an external operator, a service contractor has no authority over budgeting, staffing and implementation decisions.

Utility performance remains sole responsibility of the existing management, which may weaken the enforcement of operational changes suggested by a service contractor (OECD, 2009). This is considered the main challenge for a successful service contract: managerial commitment to its performance targets and a cooperative corporate culture towards performance management.

13.2 THE DESIGN OF THE PBC

13.2.1 Overall nature of the PBC

In 2006, Water Boards and EIB, under the ACP-EU Water Facility (the ACP-EU Water Facility was set up in 2004 with the principal objective of providing water and basic sanitation to the poor and improving water management governance in the African, Caribbean and Pacific Group of States) developed a comprehensive package of (1) a service contract to improve the utilities' financial and operational performance and oversee rehabilitation works, and (2) an investment component (including extension of WASH services to LIAs). In 2007, the Water Boards signed the 50% grant and 50% (33-year concessional) loan agreements totalling €32 million. Water Boards added €0.5 million each. The EIB disbursed the finances in tranches to the MoF, which then transferred the requested funds to the Water Boards' separate commercial accounts. The MIWD was the overseeing institution.

13.2.2 What is the PBC about?

The largest component of the EIB package was to finance €22 million of infrastructure works for LWB and BWB (Table 13.2) to mitigate water supply disruptions, to address urgent leaks and plant repairs, raise production capacity and extend WASH services to 723,000 people in Blantyre and Lilongwe's LIAs.

Table 13.2 Investment budget for the service contracts in Blantyre and Lilongwe (in million €).

Works	Lilongwe	Blantyre
Improve water production capacity and water supply efficiency	5.3	6.8
Reduction in NRW	1.1	3.6
Extension of water supply in LIAs	2.1	2.9
Total investment component	**8.5**	**13.3**

Source: EU/EIB Private Sector Service Contract.

€6 million were set aside to finance the procurement of: a transaction advisor to design the contract and assist in the tendering process, the service contactor's remuneration, and the services of an Independent Auditor to monitor the performance of service contractor. In 2007, the transaction advisor spent over a year to determine the specific features of the service contractor. The resulting contractual arrangement can be described as a service contract (with base fee) combined with performance-based elements (Figure 13.1).

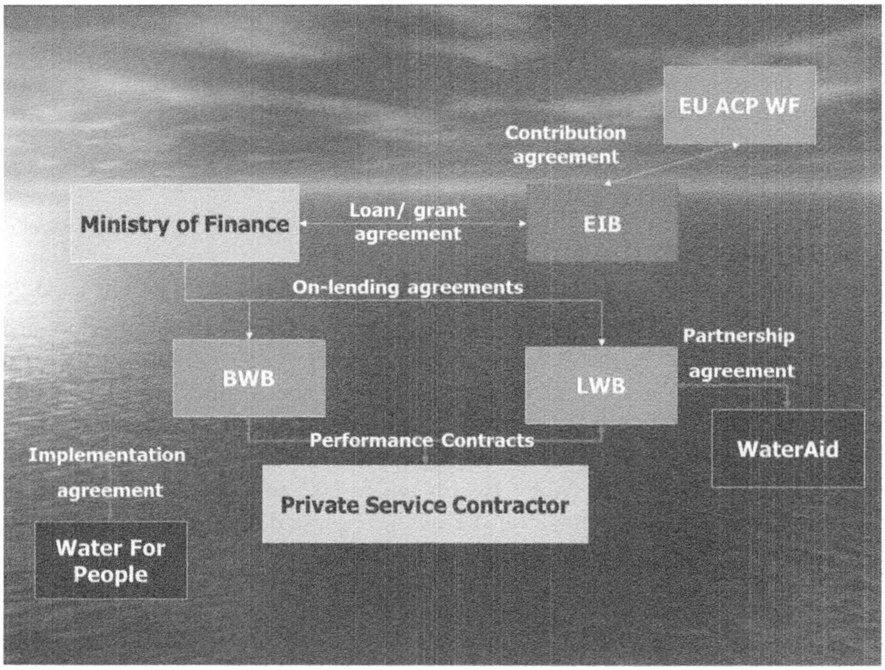

Figure 13.1 Contractual arrangements within the service contract.

The service contract was comprehensive and innovative in blending technical assistance and investments. The objectives of the service contract were to:

- **Make greater volumes of water available** through loss reduction, water demand management, rehabilitation of existing infrastructure, and development of new water resources;
- **Financially stabilise the Water Boards** through more efficient management, NRW reduction, and expansion of water sales;
- Identify and manage appropriate and affordable **WASH services for low-income areas** in a participatory approach with NGOs and local communities.

Key performance indicators (KPIs) were identified, monitored, and mathematically transferred into remuneration by the Independent Auditor. The KPIs of the service contract were not aligned to those imposed by the Government on Water Boards. Alignment of corporate and contractual KPIs is most instrumental in sharing ambitions, raising commitment, and streamlining intervention programmes.

13.3 PERFORMANCE CONTRACT DESIGN

The four-year performance-based SC had the possibility of a 12-month extension. The objectives, as described before, for both contracts with LWB and BWB were nearly identical. The SC covered a wide range of services linked to a broad range of functions within the utility that support the contractual objectives; ranging from reducing NRW to creating a coordinating platform for water kiosks in LIAs and launching an HIV/AIDS preventive program.

13.3.1 Services to be provided

The main objective was to improve the utilities' operational performance and make water boards more "bankable". The service contractor was expected to be an agent of change and recommend ways to enhance the overall management, addressing organisational and staffing structure, and reduce NRW. Results were to be achieved from a bottom-up approach: capacitating staff to re-design operational processes and act accordingly. Transparency and individual accountability was considered instrumental for all with the aim to enhance operational performance. For management this approach received lip service support but their actual cooperative support to enforce change took more than a year.

NRW is a key indicator for a well performing utility. The SC procured equipment (flow meters, clamp on meters) and systems (GPS, GIS) to reduce NRW, and developed an approach to assign responsibilities and accountability through their caretaker approach to tackle physical and administrative losses. Comprehensive staff training was initiated on accounting, customer services, customer database, leak detection and repair, GIS mapping, hydraulic modelling, operation and maintenance, etc.

The service contractor was also to supervise the investment programme, including works to improve water production capacity and the extension of water supply in LIAs, via extending the network and constructing water kiosks. The SC assisted the water boards in procuring, coordination, and supervision of the works, and provided a roadmap to manage this complex task.

Finally, the SC was to propose measures to improve the water boards' financial performance, through developing a water tariff policy, proposing cost reductions, and improving revenue collection. The service contractor also had to propose a communication strategy for the water boards (including information campaigns) to improve their customer relations and "polish their image".

13.3.2 The remuneration structure

The service contractor's remuneration had two components: a total base fee for both cities and a substantive performance-based compensation (PBC) determined by the level of achievement of performance targets. In Blantyre the level of assistance required from the service contractor was more substantive, in particular for the reduction of NRW and the supervision of the investment programme. The PBC represented 30% of the total claimable remuneration which was quite substantive. The total fees are given in Table 13.3.

Table 13.3 The total base fee of the SC.

Annual Base Fee	LWB	BWB
Contract Year 1	€600,000	€900,000
Contract Year 2	€440,000	€710,000
Contract Year 3	€250,000	€350,000
Contract Year 4	€250,000	€350,000
Total base fee	**€1,540,000**	**€2,310,000**
PBC component	**€660,000**	**€990,000**
Penalty component	**€450,000**	**€650,000**
Aggregate Fee	**€3.850,000 (+€1.650,000 PBC)**	

13.3.3 Performance regime: a "stick and carrot" approach

By a "stick and carrot" approach the SC was subject to *penalties* if the deliverables were not submitted or approved in a timely fashion, while it could perceive *bonuses* if specific performance targets were achieved. Penalties were applied to deliverables that were under the service contractor's full command, i.e. the production of action plans. By contrast, bonuses were applied to KPIs over which the SC could not entirely be held responsible.

13.3.3.1 Penalties to incentivise timely and quality services

Penalties were to be applied if the SC did not abide to the timely and quality requirements of deliverables set in the contract. To determine the penalties' amount, the contract associated *performance points* with each specific task. The contracts had particularly high reporting requirements: 76 performance points, representing nearly 50% of the total number of performance points, were associated with the production of these reports (Figure 13.2). These written deliverables came in addition to obligatory quarterly and annual progress reports and other forms of reports, including a baseline report, annual KPI reports and project completion reports. There were no penalties associated with the KPIs.

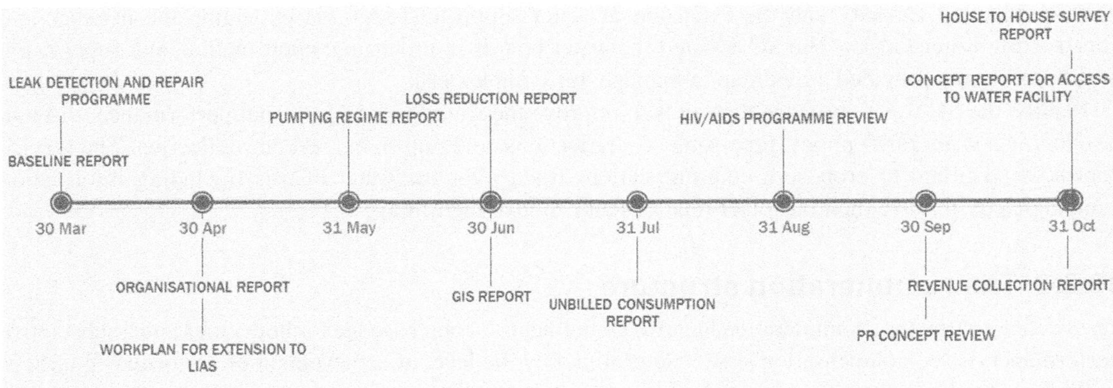

Figure 13.2 Reporting requirements in the first 12 months of the SC.
Source: Private Sector Service Contract.

13.3.3.2 Bonuses to enhance the service contractor's performance

The service contractor was incentivised with a performance-based remuneration upon achieving performance targets. The contract included four key performance indicators (KPIs): non-revenue water (NRW), working ratio (WR), continuity of supply, and supply to LIAs. BWB had one additional KPI: increasing water production capacity. Each KPI was associated with a weight for calculating the remuneration (Table 13.4).

The service contractor's performance was measured against the KPIs' values at baseline and the targets set for each SC year, as set out in Table 13.5. The service contractor was required to revise preliminary baseline values and submit a baseline report within 6 months of the contract commencement date.

The performance targets of this contract were ambitious and reflected the eagerness of the EIB to see a rapid improvement of the utilities' performance. However, with a role limited to that of an advisor, the actual direct influencing potential of the SC to sort out impact on NRW and WR was quite limited. The impact of the SC did solely depend on convincing water boards to cooperate with the SC. This implied the need for a good personal interrelationship ("chemistry") between the main actors.

Table 13.4 KPIs and associated weights.

KPIs	Units	Method of Calculation	Weight	
			Lilongwe	Blantyre
Non-revenue water	%	NRW (comprising apparent and real losses) calculated as: system input volume in m^3 – Revenue Water in m^3	40%	40%
Working Ratio	Ratio	Revenues from Sales/operation expenses for water sales (excl. depreciation) in million kwacha	15%	10%
Continuity of supply	%	Number of hours when the system is pressurised (with a minimum pressure of 1.5 bar).	30%	20%
Supply to low income areas	Number of kiosks	Number of kiosks newly installed and operating	15%	10%
Water production capacity	m^3/day	Based on test results following the establishment of new equipment	n.a	20%

Source: Private Sector Service Contracts.

Table 13.5 Preliminary performance targets and targets for year 4 (end of contracts).

KPIs	Lilongwe		Blantyre	
	Base Year	Year 4	Base Year	Year 4
Non-revenue water (NRW)	31%	25%	45%	25%
Working ratio (WR)	1.13	1.35	1.0	1.35
Continuity of supply	*To be determined by the contractor as part of a baseline report*			
Supply to LIAs	0	372	0	363
Production capacity (m^3/day)	Not applicable	Not applicable	86,000	105,000

13.3.3.3 An independent auditor to monitor performance

The IA assisted BWB, LWB and the service contractor in developing the methodology for calculating the performance achievement of the service contractor. The service contractor produced annual KPI reports which had to be approved by the water boards who would then send it to the IA, for result verification (against the baseline value and performance targets) and calculation of the performance-based remuneration (i.e. the bonus) payable to the service contractor for a given year.

13.4 THE PROCUREMENT/TENDER PROCESS

The procurement procedure was an international restricted tender procedure and involved two competitive stages: a *pre-qualification* and a *proposal stage*. The pre-qualification stage selected two bidders that would then develop full proposals. Into the bidding process a 9-week period was included during which the pre-selected bidders could conduct site visits to develop their implementation plans (including prioritisation of investments).

Due to the innovative (and risky) procurement approach water boards required that the two bidders submit a Proposal Development Bond of €250,000 to ensure their ongoing commitment to the tender process. In addition, the successful bidder was required to provide to each water board a performance security bond of €200,000 provided by a Malawian financial institution. This performance security would act as a guarantee of the proper and timely performance of the service contractor's obligations. Procurement was led by water boards (November 2007). The Internal Procurement Committees (IPCs) of BWB and LWB appointed an Evaluation Committee to evaluate prospective bidders and select the winning bid. An external party supported the procurement process and guided the water boards towards the selection of the winning bid and the signing of the final contracts.

13.4.1 The pre-qualification stage

In October 2008, a Request for Qualification (RFQ) was issued to "utilities and operators" providing an overview of the nature of the contract (duration and services), the selection process and the schedule. Prospective bidders were invited to participate in a pre-qualification meeting in Blantyre held in November 2008. Six bidders submitted Statements of Qualification to the Evaluation Committee: WSSA (South Africa), Berlin Wasser International (Germany), Rand Water (South Africa), Saur (France), Vitens Evides International (Netherlands), and Veolia (France).

The Evaluation Committee ranked the Statements of Qualification according to the following set of criteria: experience in managing, operating and maintaining water supply systems (including experience in managing water systems outside the applicant's own service area), the proposed approach to the Site Investigation in preparation for the proposal (comprehensiveness, level of details, relevance of methodology, work plan), and the financial position (an average turnover of at least €50 million was necessary to pass the test). In January 2009, the water boards officially requested VEI and Veolia to submit a full proposal.

13.4.2 The proposal stage

Both bidders signed a Proposal Development Contract, which set the terms of the proposal development stage and included a €50,000 financial compensation to the losing bidder. Bidders carried out two site visits (1 month each, separately) to prepare their approach. During those visits, bidders conducted measurements to evaluate the supply situation in the water boards' networks and assessed the proposed KPIs. Other than technical, these site visits were a good opportunity to familiarise themselves with the water boards and test their potential for future collaboration. However, the lack of data provided by both water boards reduced its effectiveness. For example, VEI was not able to gather a full set of data on flows and pressures, enabling them to assess the starting point for "continuity of supply".

The bidders submitted their financial and technical bids in April 2009. The water boards reviewed and scored the technical proposals before convening the bidders to the opening of the financial bids. VEI was officially announced as the successful bidder in June 2009. Their preparation of the proposal and development of their approach impressed the Evaluation Committee: VEI had made a substantial effort in understanding the situation on the ground. Its technical proposal was by far the best (Table 13.6).

Table 13.6 Bidding results and computation method.

	Technical Proposal	Financial Bid
VEI	83.29	3.85 Million
Veolia	74.68	3.80 Million

Final evaluation was based on the following formula: $0.9 \times TP + 0.1 \times FB$

The VEI bid included a full compensation of 8 man-years of resident staff and the services of approximately 300 man-weeks of Short-Term Expert input: 120 for Lilongwe and 170 for Blantyre from the base fee. Contracts with water boards were signed on 30 July 2009 and effectuated on 1 November 2009. VEI sub-contracted Royal Haskoning DHV for the engineering, procurement and construct supervision of the investment programme. A Dutch NGO Simavi guided WaterAid and Water for People in the extension of WASH services to LIAs including the establishment of WUAs.

13.5 IMPLEMENTATION AND RESULTS ACHIEVED

Contract implementation started in November 2009. There was gradual but accelerating progress in achieving performance targets. There was a contractual obligation to deliver action plans, to initiate procurement of goods, and to establish a solid baseline for the KPIs as specified in the contract terms. Water boards paid little attention to the service contractor's action plans; their approval was often "by default". Dissatisfaction was mounting on both sides: VEI felt uncomfortable with the "consultant" position they were confined to by the reporting obligations of the SC (with serious penalties), while water boards missed progress on the ground and towards achieving KPIs.

13.5.1 The "office" phase

In the first 18 months, the SC focussed on the production of deliverables to avoid penalties; only 25% of their input was related to "operations" (Table 13.7). The Mangochi workshop in early 2011 was a turning point. Water boards expressed their discontent with the "meagre" performance of the service contractor in achieving the KPIs, while VEI questioned the commitment of water boards. Parties agreed to accelerate field operations and reduce focus on written deliverables; VEI pre-financed an extra 60 man-weeks expert input to bring about change in operations. A proposal from VEI to water boards to take part in the spending of the "to be acquired PBC" was rejected. The service contractor was held solely accountable for achieving KPIs according to the contract. Management development programs were enrolled to strengthen coordination, alignment, and communication within boards, increase their commitment to change, and assign sufficient and capable staff to address various KPI action plans.

Table 13.7 STE inputs provided by the service contractor (in man-weeks).

STE Inputs	Writing Phase 2009–2010	Implementation Phase 2010–2013	Total
LWB	50 for writing 16 in the field	74 in the field	140
BWB	50 for writing 20 in the field	100 in the field 40 engineering support	210

13.5.2 The "operations" phase

13.5.2.1 NRW reduction

The Water boards adopted the "caretaker approach", i.e. by fragmenting the supply areas in smaller DMAs and assigning responsibility and accountability, to effectively monitor water losses. The caretaker oversees about 3000 connections and is responsible for all distribution activities within his DMA, including leakage detection and repair. Caretakers, seconded by plumbers and zones accountants, became decentralised combat units to reduce NRW. This was a massive shift for the centralised culture of the water boards and implied substantial reorganisation of the utilities' operations, including the recruitment of staff. In Blantyre, the approach started in September 2010; LWB took longer to adopt the strategy (as of January 2012).

The NRW task force formed decentralised NRW teams headed by a Zone Manager and comprised caretakers, and employees of the Customer Services, Service and Repair, GIS, Network Audit, and Metering Sections. The introduction of these zone teams created a competitive environment stimulating employees "to go the extra mile" in reducing NRW. Through on-the-job training, field employees were upgraded in effectively carrying out their assignments.

Water boards achieved slow but gradual progress in reducing NRW, although it did not reach the ambitious targets. By Year 5, NRW (12-month average) for BWB was 38% against a baseline of 48% and a target of 25%. In Lilongwe, NRW stood at 34% against a baseline of 39% and a target of 25%. According to recent data, LWB has been hitting a monthly target of 30% in October 2014, while in Blantyre NRW reached a monthly 32% in April 2014.

13.5.2.2 Working ratio (WR)

The SC stimulated a performance drive within WBs, with more efficient structures, interactive budgeting, and a new tariff calculation model. Overall, the aim was to make utilities more bankable and raise their creditworthiness. In Blantyre, the implementation gradually took of; in Lilongwe the planned merger of LWB with the Central Water Board was used as an excuse not to introduce any major change. With new General and Financial Managers on board, implementation accelerated. The new tariff calculation model that would reflect the full cost was never implemented by water boards neither did they bring it to the attention of the GoM for consideration. Both water boards fully complied with the targeted working ratio. These results indicate that both water boards improved their operational efficiency and reinforced their financial position. This is a major achievement, realising the huge investment challenges to meet the steep rising water demand projections.

13.5.2.3 Water kiosks

The Kiosk Management Units were sensitised to effectively cooperate with NGOs, City Council and customers in identification of water kiosk sites, planning their connection, establishing Water User Associations, and ensuring proper operation and caretaking. Realisation of the total of 600 water kiosks was timely in Blantyre and slow in Lilongwe. At the end of the contract period 75% of the target was realised. However, due to the perseverant water shortage in Blantyre, approximately 35 water kiosks have still not received, water simply because it is not (yet) available. With the anticipated rise in production capacity this will be remediated.

The extension of Water Supply to LIAs has been achieved. All scheduled water kiosks were planned and constructed. Crucial to this success was the partnering with WaterAid and Water for People (both NGOs) that were very instrumental to extend services, while the water boards had little means and manpower to operate or supervise the works. Most unfortunate was that after closure of the project a dispute did arise on the actual number of kiosks that were designed, constructed and operated in Blantyre during the project. Procurement is a fragile and often underscored process where lack of integrity creates "opportunities".

13.5.2.4 Rehabilitation of infrastructure

The World Bank financed design documents for rehabilitation of infra works at BWB and LWB, drafted by Posch & Partners Consulting Engineers (P&P) were according to service contractors "inadequate and incomplete" and would involve significant risks including possible quality loss, delays during works execution, and subsequent cost escalation. A proposal to re-design and prepare new tender documents (outside the scope of services) was rejected by water boards and did not enhance the relationship with Blantyre Water Board; the latter interpreted it as an attempt to generate "additional services".

The contractor recruited in Lilongwe showed excellent performance. The works were rather standard and the quality of services provided was well supervised and executed in a timely fashion. The works in Blantyre were more complex by nature and the initially recruited lowest bidding contractor Gammon Technofab Joint Venture from India was unable to deliver services. Although the contract commenced in September 2011, GTJV failed to comply with contract requirements. The contract expired in February 2013 and was not extended. GTJV had already manufactured the goods, but completion of the works at Walker's Ferry had to be retendered. The newly appointed contractor, APE commenced in September 2013 and is scheduled to finish by March 2015.

13.5.2.5 Continuity of supply

Logistics and finances constrained the use of the "continuity of supply" KPI. There was no baseline and it would have involved substantial investments in flow and pressure measurements. In full agreement with both water boards and the Independent Auditor, the associated remuneration was transferred to NRW as a basic indicator for efficient water operations.

13.5.3 The extension phase

Although the role of the contractor was "to assist" the water board with their operations, management initially failed to create ownership. Employees expected VEI to "do the job" instead of, and not with, them. Hampering this was that 3 years had elapsed between the inception of the EIB–EU project and the arrival of the contractor. Management staff involved in the design of the contract had left the water boards. Only in the operations did employees gradually recognise VEI experts as "one of them". In addition, project management teams (PMTs) witnessed VEI as a water operator in The Netherlands via exposure visits. Confident that operational performance would further improve, water boards and VEI mutually agreed to extend the SC for one year and "harvest and sustain" what was initiated. With EU/EIB approval, parties continued working on KPIs, established a mid-term investment plan, and conducted management development programs to transform Water Boards into more business oriented utilities.

13.5.4 Remuneration

No penalties were incurred to VEI; deliverables were submitted in a timely fashion and mostly approved by default. VEI increasingly with time perceived performance-based compensation (Figure 13.3). The total awarded PBC of €760,000 over its 4 years, represented just half of its maximum of €1,650.000. In order to sustain the changes introduced, VEI, in close consultation with both water boards, offered to re-invest this awarded PBC in additional technical assistance by extending the project contract by one more year. In this extension period VEI and boards managed to professionally comply with the contractual KPIs of the service contract although this was not translated into additional performance based remuneration. Most of what was initiated within the 4-year time frame of the service contract could be consolidated and embedded within the boards in this extension period in order to sustain the achievements.

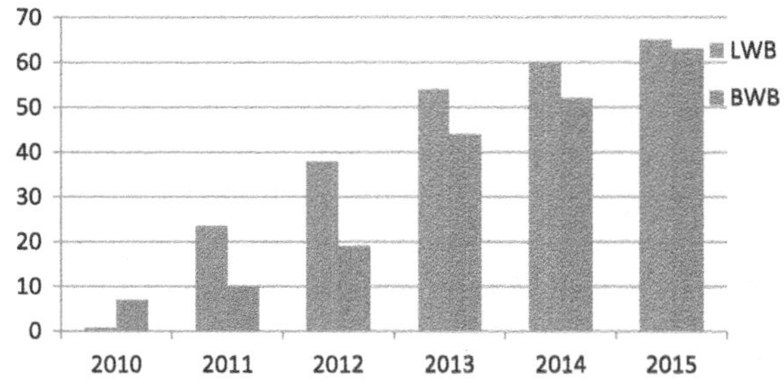

Figure 13.3 Performance-based compensation to the service contractor (in % of the maximum compensation achievable over the 5-year SC (2010–2014) and the projections for 2015.

13.6 DID THE CONTRACT ACHIEVE AND SUSTAIN RESULTS?

There will definitely be a lasting impact within Lilongwe Water Board. Through the service contractor's interventions, LWB is with its new management on the roadmap for change, realising a gradual but steady improvement in corporate culture and operational performance. Continued support in capacity and management development helps to sustain these achievements. In mutual consent, LWB and VEI have embarked on a self- financed continuation in a so-called "performance based WOP". Due to managerial turbulence within BWB, no follow-up is foreseen at short notice.

13.6.1 Internalities

The fact that the contract endured, was adjusted in 2011, and extended in 2013 by mutual agreement, clearly indicates that a strong relationship eventually took shape. Water boards appreciated the operational partnership, while personal and professional relations based on trust and accountability improved. Good relations turn out to be decisive in the overall success of a contract.

The SC has realised substantial improvements within the operations, organisation, finances, management, and last but not least, corporate culture of water boards. The appointment of a new General Manager in Lilongwe (2014) boosted cooperation. Management is convinced that it would work well for LWB to continue. In Blantyre however, the forced "leave" of its Chief Executive created a management vacuum that tended to jeopardise improvements realised within the Board.

The water boards together with the service contractor were able to realise 100% of the targeted water kiosks construction, connection and operation. The working ratios improved even beyond the set targets indicating the strengthening of their financial sustainability. The set NRW target of 25% could not be achieved in both water boards. However, there is a clear and consistent downward trend observed in 12-month average NRW figures for Lilongwe and Blantyre. They may soon hit the 30%.

The SC did bring significant changes in the corporate culture of the Water Boards, including in employees' attitude towards individual accountability and responsibility. These are difficult to quantify and may take time to yield tangible results on the ground. The end-of-contract service contractor remunerated 10 of the most cooperative employees to reward their courage and drive to team up with the SC during the service contract.

13.6.2 Externalities

Time consuming procurement procedures, customs, rising energy cost, frequent power cuts, and incidental reluctance of the GoM to part with hard currency, affected the progress of the contract. In addition, in May

2012, the Reserve Bank of Malawi devalued the kwacha by 50% (until October 2013), which increased the utilities' expenses for chemicals and electricity, impacting the WR.

13.7 REFLECTIONS ON PERFORMANCE-BASED SERVICE CONTRACTING

1. The innovative and powerful blending of investment funds with technical assistance in the contract recognises the linkage between investments and operations to improve operational performance and enhance financial sustainability. Credits go to the EIB for recognising and pasting this into the contract. Asset management principles based on life-cycle costing and sound business planning within water utilities is urgently needed to improve their financial autonomy.
2. While management contracts are a more comprehensive and strongly mandated contract for managing all or major parts of operations, in this service contract the service contactor was insufficiently mandated. Alignment of contract KPIs to corporate goals and imposing an incentive scheme to both, water board and service contractor, could enrich final contract design and yield even more results.
3. To realise organisational transformation and obtain sizable operational improvements within SCs, water utilities are to be fully engaged (and forced?) to actively take part in this – time and effort consuming – change process. If the service contract is experienced as "being imposed upon us" in order to obtain access to investment funds, it is most likely not to obtain targeted tangible results within the contract period.
4. The innovative two-stage bidding procedure, allowing 2 best bidders site investigations, provided not only an opportunity to better understand the underlying challenges of the water boards, it also recognised the importance and relevance of tasting corporate culture and building relationships.
5. Service contracts are to maximise operational flexibility in their approach rather than be too detailed and descriptive in what to do. To yield tangible results its activities automatically "engage in field operations" to achieve the targeted KPIs and its related remuneration (penalties or bonuses).
6. The "stick and carrot" approach: penalties for targets under the service contractor's full command (submission of deliverables) and bonuses to areas for which the service contractor could not be held fully responsible (KPIs) did fit VEI as not-for-profit service contractor (being more averse to penalties than attracted by profit/bonuses).
7. Questions arise as to whom to incentivise in a contractual relation. The service contractor is by default triggered to build a professional track record on what has been achieved. The recipient utility may better be challenged with "stick and carrot" by providing tapered conditional access to investment funds based on achieving tangible results in pre-defined KPIs.
8. To monitor progress, regulate communication, diffuse early tensions, and clarify expectations, external "marriage counselling" is instrumental in establishing a climate of collaboration, mutual understanding and trust. Government, the financer or an active independent assessor could take this role.
9. Retaining capacitated staff and preventing a brain drain is to be encouraged via structural performance appraisals, and corresponding remuneration and career paths. This is instrumental in bringing about organisational change and transformation of corporate culture.
10. During the project, parallel IDA/WB investment projects took place in Lilongwe and Blantyre to rehabilitate reservoirs, pump stations and treatment works. Coordination and communication between IFIs, donors, recipient government and water boards are essential to maximise anticipated results.

Finally, the gradual transformation of a “contractual arranged marriage” specifying the rules of engagement into a “confidence-based partnership” based on shared ambitions and affection, turned out to feel more natural and comfortable. It even formed the basis of a new performance-based partnership project with well-defined KPIs.

13.8 ACKNOWLEDGEMENTS

The enabling and innovative environment of the private sector service contract (PSSC) in Malawi was created by the European Investment Bank and the European Union. The Contracts between Vitens Evides International (VEI) with Lilongwe and Blantyre Water Boards were comprehensive in blending technical assistance with investment components, while it promoted active participation of local NGOs to provide supporting services in LIAs. I am grateful to all staff of LWB and BWB for their valuable contributions, and to VEI resident project managers and short-term experts for their everlasting effort to turn this Performance-Based Contract into a professional success.

13.9 REFERENCES

Cammack D. (2012). Peri-Urban Governance and the Delivery of Public Goods in Malawi, 2009–2011. O. D. Institute (ed.), Africa Power and Politics Programme (DFID & Irish Aid). Published on behalf of the Africa Power and Politics Programme (APPP) by the Overseas Development Institute, 111 Westminster Bridge Road, London SE1 7JD, UK (www.odi.org.uk)

EIB (2007). Appraisal Report: Malawi Peri-Urban Water and Sanitation. Unpublished.

IWA’s Task Group on Performance-based Contracting (2011). Performance-Based Contracting: A Mechanism for Increasing Water Utility Efficiency. IWA, London.

Kingdom B., Liemberger R. & Marin P. (2006). The Challenge of Reducing Non-revenue Water (NRW) in Developing Countries. How Can the Private Sector Help: A Look at Performance-based Service Contracting. PPIAF, The World Bank, Washington DC.

Ministry of Irrigation and Water Development (2007). National Water Development Programme.

OECD (2009). Private Sector Participation in Water Infrastructure – OECD Checklist for Public Action. OECD.

Trémolet S. (2006). Case Study on Senegal’s Water and Sanitation Sector Economic Regulation. The World Bank, Washington DC.

Veenstra S. (2014). Teamwork Works. Illustrated documentary of the Peri Urban Service Contracts with Lilongwe and Blantyre Water Boards, VEI, The Netherlands.

WaterAid (2012). Managing Communal Water Kiosks in Malawi: Experiences in Water Supply Management in Poor Urban Settlements in Lilongwe. WaterAid, London.

WSP, UN-Habitat, ESAR-IWA, AAE (2009). Water Operators Partnerships, Africa Utility Performance Assessment. WSP, Nairobi.

Zeleza Manda M. (2009). Water and Sanitation in Urban Malawi: Can the Millennium Development Goals be Met? A Study of Informal Settlements in Three Cities. International Institute for Environment and Development IIED, London.

Chapter 14

Performance-based affermage contracts

Vivian Castro-Woolridge, Didier Carron and Jan Janssens

Affermage contracts are widely used in the water sector by water utilities, especially in France, in order to delegate operation and maintenance to a private contractor. While technical performance is often – but not always – provided by the contractors through such contracts, several drawbacks had become visible over the years, especially because the profit made by the contractor is not correlated to its performance. Performance-based affermage (PBA) contracts make it possible to relate the profit resulting from the private operation to the performance of the contractor and change the earning logic, as they can ensure an incentive-driven and more equitable distribution of the financial surplus between the operator and the water authority. This profit sharing can be based both on technical performance indicators such as customer service, technical performance or sustainable development and on financial indicators such as productivity improvement. Return on experience of the SEDIF case (France) shows how effective such PBA contracts can be.

14.1 THE BASICS OF THE AFFERMAGE CONTRACT

An *affermage* is one type of delegated management contract in the private–public partnership (PPP) spectrum which is widely used throughout the world, and especially in France. Under this type of contract, the operator is responsible for operations and maintenance. The operator collects the tariff directly from consumers on behalf of the contracting authority (CA). The CA is usually responsible for major rehabilitation and new capital works. However, each contract defines the exact terms and responsibilities for financing and implementing maintenance, rehabilitation and new works.

The ultimate responsibility of delivering the service to the customer however remains with the CA, which then must constantly control the service by the operator in order to make on time all key decisions regarding the service.

Authors: Jan Janssens, Didier Carron, Philippe Marin, Philip Giantris and Tom Williams
doi: 10.2166/9781780405964_0135

In water management the tariff collected by the private operator typically includes three components:

- A fixed fee per connection + a fee per cubic metre of water sold which are kept by the operator as a counterpart to its services. This fee is the price for the operator's services defined in the affermage contract.
- A fee per cubic metre which is paid to the CA and forms the main revenue of the water budget of the CA.
- Taxes, typically to the regional water agency and to the State (tax on added value).

The *affermage* combines public financing with attracting private efficiency. It transfers most of the technical and the commercial risk to the operator. It benefits from know-how and experience of private operators. Heavy investment remains with the CA, which can typically get cheaper money from public financial institutions than private companies.

14.2 DIFFERENCES AND SIMILARITIES BETWEEN THE AFFERMAGE AND OTHER OPERATIONS CONTRACTS

Affermage, originally a French word, is often translated in English as "lease contract" or "management contract".

However, these terms address different types of contracts. Table 14.1 describes a few of the key differences and similarities between the affermage contract model, the lease contract model and the management contract model.

Table 14.1 Key features of the affermage, lease and management models.

Features	Affermage Contract	Lease Contract	Management Contract
Mission	Operation including OEM	Operation including OEM	Co-management
Operator's revenue	Typically based on €/m^3 sold (*volumetric*) from customers	Annual monetary from utility	Fixed fee (typically 70%) (annual monetary) + variable fee on performance
Competitive bidding process	Lowest water tariff proposed wins *The operator's price covers O&M costs, including some renewal costs*	Highest bid (lease fee) wins *The lease fee is paid to the Contracting Authority by the operator*	Lowest fixed fee and highest commitments on performance wins *Both fees are paid to the operator by the CA*
Performance incentives (when any)	Linked to technical and financial performance (m^3)	When any, typically linked to water sales	Linked to technical and commercial performance (m^3)

14.3 AFFERMAGE CONTRACTS FROM THE 19TH CENTURY UNTIL TODAY IN FRANCE

As described by Lorrain (2008), affermage in the water sector originated in 1853 with the creation of the Compagnie Générale des Eaux (CGE) by an imperial decree. The CGE obtained in Lyon (France) its first concession of public water distribution service. Seven years later, Paris entrusted to CGE its water distribution service for a period of fifty years.

Since then, affermage progressively differentiated from concession with the emergence of public local power and the huge amount of investment needed to build major water infrastructures, which could only be provided by public money.

In the 80s and early 90s, many French municipalities decided to entrust the operation of water services to one of the three French operators, namely, Compagnie Générale des Eaux, now Veolia, Lyonnaise des Eaux, now Suez, and Saur. This move was pushed by the impressive know how accumulated by these companies, both from a technical point of view but also regarding billing, collection and customer relations. However a less worthy reason why affermage was also selected is that it could offer a way to finance other infrastructures via an entry fee or even, in an illegal and hidden manner, political parties.

In 1991, a law was passed making compulsory a public tender for every new contract, to create emulation and prevent corruption. Entry fees were banned and duration of contracts became limited to 20 years except when large infrastructures were financed though the contract. In 2006, 72% of the French population was supplied by a private operator. In the decade starting in 2000, experience of about 15 years was thus available over this new wave of contracts.

While technical performance was often – but not always – provided, the following less pleasant features were often visible:

- The risk on volumes, main risk for the operator as its revenue is mainly proportional to volumes sold, never materialized in the context of increasing consumption due to urban development. On the contrary, these increasing net sales generated high revenues with high profits.
- Price revision formulas did not take increase in productivity into account, resulting in excessive increases of the tariff of water reinforcing high profits by the operators.
- Water authorities felt a loss of control over the service, partly due to lack of information delivered by the operators.
- Poor competition was sometimes seen at the end of the contract when tenders for a new contract were launched, resulting in new contracts not well balanced between the parties.
- Asset management, which was often a grey zone in the contracts, resulted often in poor or uncertain performance, with lack of management of non-revenue water.

14.4 THE WAY TO A SECOND GENERATION OF AFFERMAGE CONTRACTS

This was the start of a strong debate in France aimed at comparing public operation and private operation of a water or wastewater service.

For those municipalities wanting to maintain a setup with a private operator, the need to improve conventional affermage contracts became clear, with a view to keeping the strengths of the model and proposing new implementation features to address weaknesses.

A major rejection of this view occurred when the largest water utility in France, the SEDIF (2011–2015), grouping 144 municipalities from the suburbs of Paris, accounting for a population of 4.3 million, decided in 2008 to stay with a PPP model for its operation at the end of the operation contract with Veolia, effective on 31 December 2010. However, the SEDIF was determined to establish a new type of affermage contract, fully correcting the weaknesses of conventional affermage contracts and relating the revenue of the operator to its performance.

The process resulted in what can be called a performance-based contract, opening the way to a second generation of affermage contracts which started to be used by many French water authorities wanting to use a private operator (Janssens *et al.* 2011).

These new types of contract include 6 main features:

- ✓ Performance-based mechanisms;
- ✓ Open book obligations;
- ✓ Mechanisms to avoid unjustified high profits;
- ✓ Creation by the operator of a dedicated company (SPV) to manage the contract;
- ✓ End of contract clauses to ensure mutability;
- ✓ Quantified asset management obligations.

Interestingly enough, some of these features were also considered during the same period in other parts of the world, as can be seen in the "Guidelines for performance-based contracts between water utilities and municipalities" (OECD, 2010).

14.5 THE PERFORMANCE-BASED AFFERMAGE (PBA) MODEL

The PBA model is based on an incentive-driven and more equitable distribution of the financial surplus between the operator and the water authority. The PBA model is also innovative for its combination of both operational and financial parameters for calculating the operator's revenue and bonus.

In a conventional *affermage* contract, the operator pays for operations and maintenance costs, remits the difference between average tariff and operator's price to the water authority, but retains the entire operational profit. In a PBA, the operational profit is shared between contracting parties according to an incentive structure that combines both operational and financial performance indicators – and which are explicitly defined in the contractual agreement.

In some PBAs, the operator's surplus is transferred to an escrow account in order to create comfort for the operator. The operator can only access these funds when he achieves both the financial and technical performance targets.

In case of poor performance vis-à-vis the contractual targets, the operator may only earn a reduced part of the surplus, or even no surplus. In case of high performance vis-à-vis the contractual targets, the operator may earn a large part of the surplus, up to 100%, depending on the contractually agreed equation for distributing the surplus between the parties.

14.6 THE EARNING LOGIC

In PBAs, the earning logic is no longer based principally on water sold (Carron *et al.* 2012).

A financial operation surplus is calculated which is the difference between the revenue (made mainly by the sale of water to consumers) and the expenses (staff, purchases, fixed fees, amortization, financial fees, if any).

The surplus is the profit of the SPV when such a company has been created to manage the contract. If the surplus is negative (i.e. the contractor has lost money), the situation remains as it is. If the surplus is positive, contrary to the conventional affermage contract, it is not fully kept by the contractor.

Only a part of this surplus is kept, as a function of the operator's performance:

- If the contractor has reached all performance targets, it can keep 100% of this surplus;
- If the contractor has not reached all performance targets, but only $X\%$ of the targets, it will keep only $X\%$ of the surplus, and pay back to the contracting authority the remaining part, i.e. $(1 - X)\ \%$ of the surplus.

In other words, PBAs include a profit sharing according to the contractor's performance.

Some specific features might alter this earning logic. Especially, a safeguard mechanism is also added, to temper potential high profit by the contractor. For example, above 9% of profit, the "excess profit" can be shared 50/50 between the contractor and the contracting authority.

The earning logic of PBAs shows numerous benefits, which include:

- **Increased value for money** – Performance risk lies with the contractor and full remuneration is contingent upon verifiable performance improvements;
- **Greater flexibility for the contractor** – Decides on measures to undertake to achieve targets, applying innovation and creative solutions;
- **Increased emphasis on accountability** – The contractor's remuneration is tied to verifiable progress vis-à-vis its targets;
- **Support to long-term improvements** – Efficiency gains through a PBA can help support long-term improvements;
- **Driver for saving water and reduce non-revenue water** – In conventional affermage, the profit of the operator is highly dependent on the amount of water sold. There is thus no drive to help consumers reduce their water consumption nor to reduce non-revenue water. In PBAs, the technical performance can include water conservation targets, and thus trigger effective actions by the operator;
- **Financial return to the customers** – The amount paid back by the contractor can be used by the Contracting Authority either for new investments, or to decrease the tariff.

14.7 TECHNICAL PERFORMANCE INDICATORS

Technical performance indicators reflect the performance of the operation regarding typically:

- Customer service,
- Technical performance,
- Sustainable development.

Water quality indicators can be included, but more generally this subject is rather subject to black or white evaluation: water quality has to meet regulations values, but does not need to overperform these values by much.

The total number of indicators can vary from a few dozen to more than 100. However not all of them need to be included when calculating the global technical performance of the contractor. Some or many of them can just give rise to penalties when a minimum value is not reached.

For indicators contributing to the calculation of the global performance level, it is preferable to use a two-threshold scheme, as described below:

For each indicator, two values are defined:

- A target value, which provides the full number of incentive points to the operator when the indicator reaches this value or overruns it;
- A minimum value, below which no incentive point is provided to the operator and which can even trigger penalty points for the operator, typically proportional to the gap between the actual value and the minimum value.

In between, a progressive incentive can be defined, in order to incentivize the operator to do its best to reach the target value.

Figure 14.1 illustrates, for a given indicator, the possible variation of the incentive as a function of the value of the indicator.

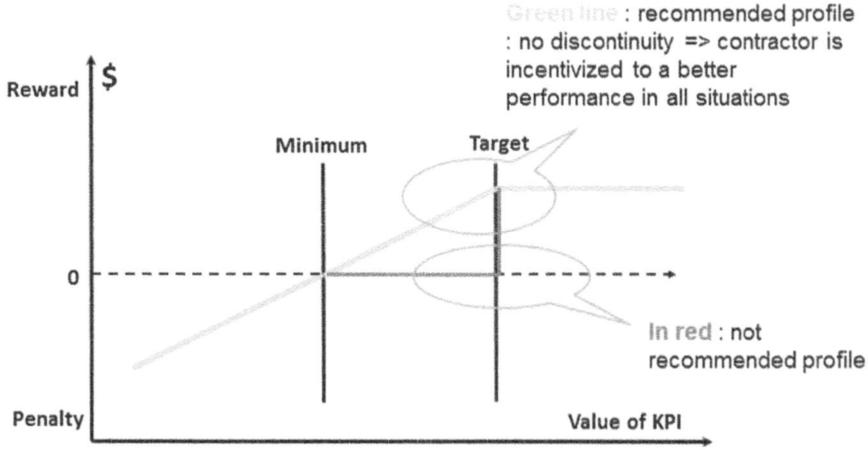

Figure 14.1 Incentive variation according to performance.

14.8 FINANCIAL PERFORMANCE INDICATORS

Financial indicators typically cover two kinds of targets:

- **Profit of the contractor**. Such a profit is essential so that the private operator finds some interest in the contract. But as the profit of the contractor is shared, it is also of interest to the contracting authority to have this profit shared.
- **Productivity improvement**. It is of general interest that throughout the year the contractor improves its productivity, mainly:
    - By taking advantage of investments typically made at the beginning of the contract and providing productivity improvements in operations;
    - By taking advantage of technology improvements available on the market (new technologies for information and telecommunications, new energy saving equipment, etc.).

Regarding productivity improvement, it can be however quite tricky to design indicators reflecting this improvement. Typically, three approaches can be used:

- Reference to professional ratios. To be fully relevant, these ratios should be adapted to each contract and specifically calculated using as input date the main characteristics of the water service;
- Reference to the previous years;
- Reference to a contractual target of expenses, with commitment of the operator defined during the tender period for awarding the contract.

14.9 CASE STUDY: THE SEDIF

The SEDIF is the water authority managing the production of potable water and its distribution water production over 149 municipalities around Paris, representing a population, of about 4.4 million (Figure 14.2).

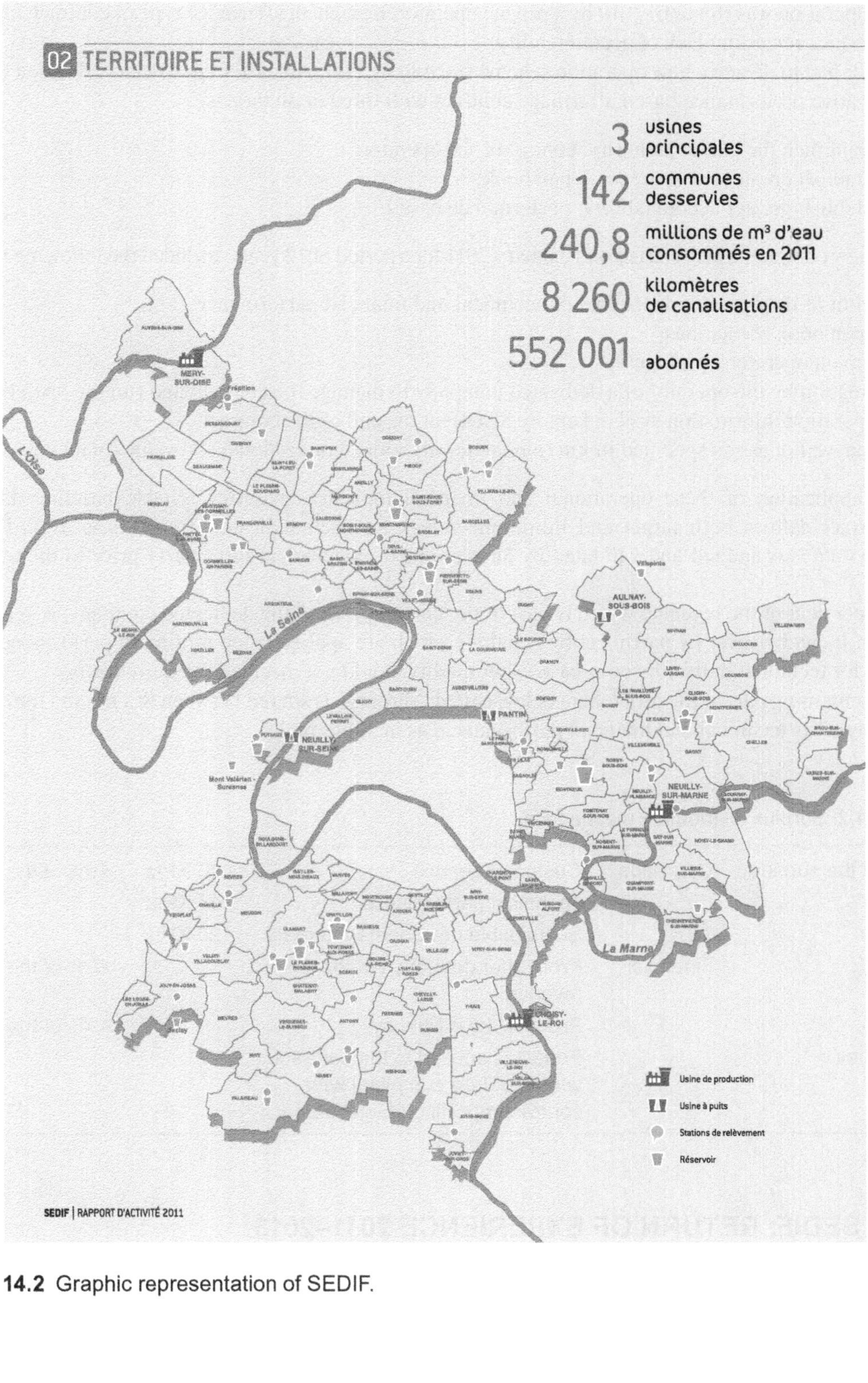

Figure 14.2 Graphic representation of SEDIF.

The operation was run until 2010 by a private operator through an affermage type of contract, and faced several critics regarding lack of accountability.

SEDIF had to decide a new operation scheme starting in 2011, and, after open debate, decided to go for an innovative performance based affermage contract with three main targets:

- strengthen the public authority: control of the operator;
- establish greater transparency: open book;
- establish greater accountability: performance based.

The new contract, which started on 1 January 2011 for a period of 12 years, included the following features:

✓ Profit of the operator dependent on technical and financial performance;
✓ Open book obligations;
✓ Cap on operator's profit at 9;
✓ Creation by the operator of a dedicated company to manage the contract and run the operation;
✓ Operation information system kept by SEDIF at the end of the contract;
✓ Renewal of assets specified in km (pipe renewal), units (connections) and value (plant equipment).

The application of these operational and financial indicators requires reliable baseline data; and the contract defines both target and minimum values for each indicator. In the case of SEDIF, the baseline data was audited and validated by an accredited independent third party prior to the tendering process.

Two per cent of the revenue, which is part of the annual profit of the dedicated company, is kept by the operator. It can however by partly cut by penalties which are applied when minimum performance is not reached for technical indicators or in case water quality would not meet legal requirements.

The remaining part of the profit, here below called "surplus", is shared between SEDIF and the operator according to its technical and financial performance, as in Table 14.2.

Table 14.2 Surplus distribution mechanism.

100% of the surplus	Operation	Customer service	16 KPIs	40% of the surplus
		Technical performance	17 KPIs	
		Sustainable development targets	20 KPIs	
	Financial	Profit after deduction of the 2% revenue		20% of the surplus
		Productivity efficiency		40% of the surplus
Open book		Penalties applied to the contractor when not fully compliant with contract reporting obligations		

14.10 SEDIF: RETURN OF EXPERIENCE 2011–2013

After 5 years of life, the technical performance of the operator has been as in Table 14.3 (SEDIF, 2011–2015).

Table 14.3 Technical performance of the operator.

	Customer Service	Operational Technical Performance	Sustainable Development	Global Operation Performance
2011	78%	81%	80%	80%
2012	78%	82%	78%	79%
2013	78%	87%	86%	84%
2014	81%	82%	89%	84%
2015	91%	70%	87%	83%

This shows both that the operator performs globally well, but also that although the target values set for the performance indicators are challenging, they are quite achievable. Only a few indicators on sustainable development have been reached since the beginning and could have been defined in a more challenging way.

Regarding more global features, return on experience after 3 years can be expressed as in Table 14.4.

Table 14.4 Return on experience.

Parameter	Outcome	Assessment
Net profit before taxes	Ranging between 4.6% and 8.9% after application of the performance-based profit sharing mechanism but before tax in profit, which can be considered as a reasonable profit appropriate to the risks borne by the operator.	
Technical and customer service performance	Between 79% and 84%, which is a high performance given that 100% is quite demanding), which demonstrates that KPI targets comply with the SMART principle.	
Open book	In place, and audited every year by an independent auditor. Full access by the SEDIF to all native data, which was a major requirement of the contract so that SEDF can get all relevant information to establish its strategy.	

14.11 CONCLUSION

Performance based affermage contracts (PBAs) have been called "second generation" affermage contracts. As a matter of fact, they appear to be a valid answer to several critics on conventional affermage contracts, especially regarding technical and financial accountability.

This contractual arrangement may be more politically attractive to public authorities, particularly in contexts where officials and civil society may feel that private operators can benefit disproportionately from operation contracts.

Return of experience show that accountability and increased value for money are effective. The decorrelation of operator's profit and the amount of water sold could be successfully implemented.

There is not one single way to define such contracts. On the contrary, they can and must be tailored to the context, be challenging for the private contractor, but also be careful to keep equitable mechanisms to remain attractive for both parties.

14.12 REFERENCES

Carron D., Janssens J. G. and Castro-Wooldridge V. (2012). Evolution of contract structures in water supply and sanitation. *Journal of Water, Sanitation and Hygiene for Development*, **3**(3), 423–431.

Janssens J. G., Carron D. and Castro-Wooldridge V. (2011). The rise of hybrid delegated management contracts: emerging lessons from France. *Water Utility Management International*, June, pp. 30–33.

Lorrain D. (2008). La naissance de l'affermage : coopérer pour exister (The birth of farming: cooperating to exist). *Entreprises et histoire*, **1**(50), 67–85.

OECD (2010). Guidelines for Performance-Based Contracts Between Water Utilities and Municipalities. Lessons learnt from Eastern Europe, Caucasus and Central Asia. OECD – 14–15 October, Almaty, Kazakhstan.

SEDIF (2011–2015). website – Résultats du contrôle de la délégation de service public (Results of the control of the public service delegation). See http://www.sedif.com/controle-de-la-delegation.aspx (last accessed 2015).

Chapter 15

Performance-based contracts, the Aroona Integrated Alliance experience

Jérôme Douziech and Donna Buchanan

Aroona Integrated Alliance is an agreement between the Water Corporation of Western Australia (Water Corporation), SUEZ and Broadspectrum to deliver water production and treatment, and wastewater treatment services to the metropolitan area of Perth (population two million). Aroona is an aboriginal word usually meaning "place of refreshing" or "place of running water". The Alliance scope is unique as it includes asset management, operations and maintenance. Aroona has 380 employees which include 260 staff seconded from the Water Corporation into the Alliance. The 10-year Alliance commenced on 1 July 2012 and is designed to harness the expertise of all three partners to achieve six service objectives. This commercial arrangement is also designed to achieve a high integration of functions and activities within the organization, as well as integration of people from multiple employing organizations that bring the full complement of capabilities required. In order to address the contractual expectations and the changing economic environment, Aroona implemented a specific strategy called "Good to Great". The "Good to Great" program imposes a set of stretch targets superior to the contractual expectations. This case study demonstrates the significant progress to date with strong safety performance and $31M cumulated sustainable savings to June 2016. It also describes the 5-year Good to Great strategy.

15.1 CASE STUDY DETAIL

15.1.1 The Aroona Integrated Alliance to deliver improvements in water and wastewater services

In the mid-1990s, the Water Corporation adopted a segregated approach to its water and wastewater services, with three specialized teams – the Water Production Branch, the Wastewater Treatment Branch and the Mechanical and Electrical Alliance. Each team focused on driving efficiencies in their service areas and addressing individual challenges. This was a strategy that was appropriate for the time and produced excellent outcomes.

Authors: Jan Janssens, Didier Carron, Philippe Marin, Philip Giantris and Tom Williams
doi: 10.2166/9781780405964_0145

During this time, the Water Corporation experienced four broad types of Alliance contracting models:

- Project Alliance;
- Program Alliance to carry out a *number of agreed projects*;
- Program Alliance to carry out projects *for a set period of time*;
- Operation and Maintenance Alliances.

In 2009, the *Integrated Alliances Project* was established to improve the efficiency and cost effectiveness of services to Perth metropolitan customers through arrangements with the private sector. Water Corporation investigations showed that the best method would be an integrated delivery model where operations, maintenance and asset management worked closer together. One option to achieve this was to partner with the private sector in the form of two alliances, one for Production and Treatment, and one for the Network. Each alliance would encompass everyone involved in delivering services to our customers under a single organization.

More precisely, the "integrated" part of the alliance comes from:

- The integration of functions and activities so that barriers or silos are reduced and single point accountability can be supported;
- The integration of people from multiple employing organizations that bring the full complement of capabilities required;
- The increased representation of Corporation in the Alliance, i.e. in leadership, management and delivery aspects.

In July 2012, the Water Corporation and its private partners, Broadspectrum and Suez, established the *Aroona Integrated Alliance* to deliver more efficient and cost-effective water and wastewater services for Perth metropolitan customers. The Water Corporation clearly articulated its expectation:

> *It is expected that this Integrated Alliance will deliver step change improvements in achieving the Service Objectives. The achievement of the Service Objectives under a commercial framework is expected to drive greater efficiency and effectiveness in the delivery of outcomes than would otherwise be achieved under the current arrangements.* (Expression of Interest, 2011.)

As a consequence, the Water Corporation expectations are that the Alliance will deliver greater efficiency while achieving six specific service objectives, which are summarized below:

- **Zero Harm:** to improve the safety performance and embed a zero harm culture;
- **Water and Wastewater Compliance:** to deliver on water and wastewater production and compliance targets;
- **Asset Management:** to improve maintenance planning, asset assessment, capability planning, asset acquisition and deliver the minor works allocated to Aroona;
- **People Capability:** to ensure capability and engagement, succession planning, knowledge transfer and knowledge retention within the Alliance;
- **Value for Money and Efficiency:** to achieve the capital and operational expenditures efficiency target in a sustainable manner and;
- **Information Management:** to provide accurate and reliable data in a timely manner.

15.1.2 A commercial framework set up to generate the right behaviour

To achieve this expectation, the Water Corporation has established a *Performance-based Agreement*, through which the Water Corporation and its private partners are collectively accountable to reach a minimum level of performance defined by a set of criteria, both on operational performance and financial performance:

- Minimum conditions of satisfaction (MCOS) defines compliance to law, regulation and the relevant Water Corporation policies;
- The balanced scorecard defines a set of key performance indicators to measure the performance delivered by the Alliance against specific contractual objectives in the six Service Objectives;
- A budget is defined on a yearly basis, including a minimum of 2% cost efficiency every year.

In addition to these necessary conditions, there is a strong incentive for the partners to generate sustainable savings: a specific *Risk/Reward mechanism* is defined, and the partners share the benefits and the risk of the Alliance performance. More specifically:

- If there isn't any breach in the MCOS, the balanced scorecard is achieved and there isn't any budget overrun, the partners are entitled to get a share of the sustainable savings achieved in the Financial Year;
- If there is a breach in the MCOS or the balanced scorecard is not achieved, the partners are not entitled to a reward;
- In case of an Alliance budget overrun, the partners will make a payment that, in any case, will not exceed the margin payable in respect of that financial year.

In setting the Alliance budget for the following financial year ($n + 1$), the Water Corporation reduces the authorized direct costs by the greater of 2% or the value of the sustainable savings achieved in the financial year (n). Through this mechanism, the Water Corporation has ensured that:

- A minimum of 2% efficiency is expected every year;
- The savings achieved in a specific financial year need to be maintained across the time, otherwise the Alliance will be exposed to a budget overrun.

As mentioned above, one-year value of the sustainable savings is shared between the Water Corporation and the private partners, if – and only if – the necessary conditions are met (MCOS and balanced scorecard, as well as no budget overrun). On the other hand, a reduction of the term of the agreement may apply if the necessary conditions are not met, and a budget overrun is shared between the partners.

This is clearly a strong driver to implement improvements in the Alliance and ensure in the meantime that the expected performance is delivered in a sustainable manner.

15.1.3 Alliance governance

Aroona has been established by the Water Corporation to operate within a well-defined governance structure that operates under the key principles of the alliancing contracting model:

- Alignment of vision/objectives/goals;
- Increase flexibility;
- Collective responsibility for delivery;
- Focus on successful outcomes for all (win/win);
- Clearly define scope, expressed outcomes;
- Share risk/reward;
- Best for alliance resourcing;
- No blame culture;
- Success is measurable and measured;
- Clear understanding of joint and individual responsibilities.

Aroona is governed by the Alliance Leadership Team (ALT) with members drawn from the three partners, with the Alliance General Manager reporting to the Chair of ALT. The ALT's role is to develop the Alliance culture and to provide direction and governance to the Alliance Management Team (AMT) led by the Alliance General Manager (AGM). The ALT appoints the AGM. The ALT in consultation with the AGM appoints the AMT on a best for Alliance basis, they are drawn from the three organizations. No doubt the strong and efficient governance is a key success factor of an Integrated Alliance, as this model introduces a cultural change that requires one to go beyond the parent/child or client/supplier contracting model.

15.1.4 The "good to great" strategy to deliver a step change

It was identified at an early stage of the Alliance (at the end of the first year) – that the significant changes in the economic environment required a step change in the Alliance strategy to make sure that value for money was delivered. For this purpose, a set of stretch goals – far beyond contractual expectations – were defined for each of the contractual six service objectives. These goals are detailed in Table 15.1.

Table 15.1 Good to Great goals.

Service Objective	Good to Great Goals (June 2018)	Measure
Zero harm	• Zero injuries (0 Lost Time Injury, 0 Medical Treatment Injury, 0 Minor Injury)	0
	• No post asset transfer safety modifications required	0
Water & wastewater services & compliance	• Compliance with Water Safety Plans and Water Production Plan	100%
	• Improvement in wastewater compliance	Increase 25%
Asset management	• Maintenance strategies reviewed and optimized at the targeted plants	31 Systems
	• All critical assets with gaps have a current mitigation plan in place	
	• System Asset Management Plans in place for all systems	
Value for money & efficiency	• Realize sustainable savings	$25M
	• Volume of capital works and external works managed by Aroona per annum	$25M
	• Deliver the balanced scorecard and minimum conditions of satisfaction	Achieved
	• Extreme and high risks have mitigation plans in place and actioned	Achieved
People capability	• Leadership capability – % in mastering or role modelling in the leadership assessment framework	80%
Information management	• Agreement on data requirements in place.	100%
	• Accurate and timely reporting via BSC and/or BPR	100%

The "Good to Great" strategy – was launched in early 2014 to achieve specific goals for each of the service objectives at the end of the 5-year program (June 2018).

Specific action plans have been developed with the workforce and the partners by utilising the three key mechanisms (shown in the circles in Figure 15.1) to develop strategies to achieve these goals. They are reviewed on a quarterly basis, and the business plan is adjusted every year taking into consideration the results achieved, existing gaps and the desired outcome and resources required.

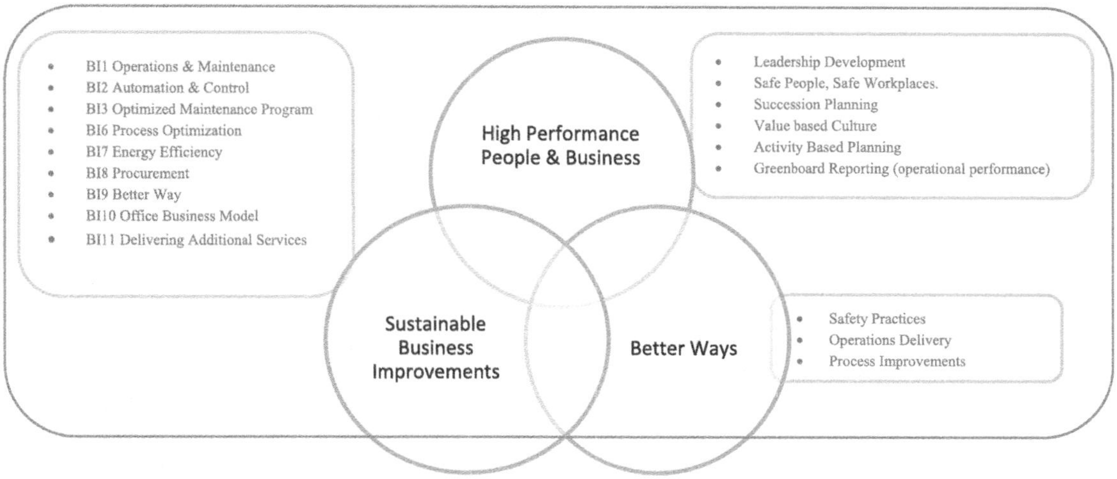

Figure 15.1 Key mechanisms used to build our Good to Great strategy.

15.1.5 A governance model is in place to ensure business improvements are sustainable

The targeted efficiencies are achieved through a *Business Improvement Program* that provides a suite of initiatives and projects across the Alliance. For an initiative to qualify as a business improvement, it must:

- Be of benefit to Aroona or in the manner of the performance of the Alliance services.
- Materially and positively impact on part of Aroona's operational performance efficiencies.
- Be *sustainable* (meaning that the benefit is expected to last unless the external conditions change).
- Be readily *measurable*: for each initiative, a baseline is defined to assess precisely the benefit of each project, and check on a regular basis over time the long-term effect of this initiative. The baseline is defined through technical and financial indicators that can be readily measured (cost, cost per unit of production, energy consumed per unit of production and quantity of active chemicals per unit of production).
- Have a high probability of a successful outcome.
- Not be subject to another incentive regime.
- Be approved by the ALT to enter the program.

A specific process has been defined and implemented to ensure that the conditions are satisfied, as well as appropriate control mechanisms are in place to control the sustainability of the efficiencies realized (Figure 15.2).

15.1.6 A Business improvement program designed to deliver sustainable cost efficiencies

As part of the Good to Great Strategy, business improvements (BIs) are initiatives intended to improve the performance and cost efficiency of Aroona's operations, maintenance and asset management. The program currently includes more than 50 projects structured into specific improvement areas:

- **BI1 Building an integrated operations & maintenance model:** BI1 program aims to improve productivity and efficiency by addressing planning, culture and knowledge as well as optimising opportunities. A suite of potential projects has been defined after a comprehensive review done in the field at the early stage of the Alliance with the workforce. For each project, the cost, potential benefits and complexity to implement have been assessed, as illustrated in Figure 15.3.
- **BI2 Automation and control:** BI2 is a suite of projects to increase the level of automation in the water and wastewater treatment plants, as well as ensuring consistency in implementing SCADA solutions on the assets. It is expected that this program will be a major enabler to realize additional opportunities in the medium term under the BI1, BI3, BI6 and BI7 programs;
- **BI3 Maintenance optimization:** a comprehensive review of the maintenance practices has been launched through a hybrid approach, combining elements of both: reliability centered maintenance (RCM) and failure mode and effects and criticality analysis (FMECA), to optimize the planned maintenance activities by targetting resources and expenditure where the greatest benefits will be realized. Pilots conducted in the early stage of the Alliance have given indications on which is the most appropriate approach that can be used. The approach involves conducting a criticality based review of each scheme. This is a high-level review that may identify either assets or modules whereby a simplified optimization or RCM approach would be the most beneficial tool to use. In addition, we have been progressively replacing the time-based maintenance practices with a performance-based practices, where it is possible to implement an appropriate real-time monitoring of the asset performance;
- **BI6 Optimizing the water and wastewater treatment processes:** this program aims to optimize the treatment processes to reduce chemicals, energy and increase the plant performance where it was necessary. This has required a comprehensive review of the treatments in place with the support of technical experts seconded by the partners, and a progressive implementation of solutions with, in some cases, pilots to assess the effectiveness of the proposed changes;
- **BI7 Energy saving and recovery:** this program includes all the initiatives aiming to reduce our energy consumption across the bores, water and wastewater treatment plants and pumping stations. It includes not only the promotion of new technologies, but more often the optimization of operations and maintenance practices. Several projects related to recovering energy from the assets under our management are targeted (microturbines for hydroelectric recovery, increasing the cogeneration performance, solar panels).
- **BI8 Procurement:** BI8 includes all the initiatives aiming to increase the value provided by contractors and suppliers, either by selecting more suitable products or by renegotiating, when appropriate, commercial arrangements in place, to leverage on the market offer. In addition, the supply chain processes are being reviewed to reduce time for delivery to the working team and ensuring an effective store management.
- **BI9 Better Way**: The Better Way program is a bottom-up innovation program that encourages employees to submit ideas to improve the business. A qualified manager is dedicated to this program to make sure that we generate an innovation momentum within the Alliance, and all ideas are followed up and, as far as practicable, can generate sustainable improvements. As of February 2017, the Better Way program had generated 1500 ideas since the inception of the Alliance, and almost 40% of them have generated sustainable improvements.
- **BI10 Head office functions:** review collates initiatives to optimize the service provided by the support functions, and more generally the head office based functions, to the core processes of the business. By reducing duplication and tasks with limited added value or reallocating resources to where they are the most profitable, we have been able to realize productivity, mostly by insourcing activities provided by external contractors;

- **BI11 Delivering additional services:** we include under this program the projects that leverage on our in-house expertise and capability to deliver additional services that are currently outside of the Alliance scope, to the Water Corporation and enable other business units within the Corporation to realize some savings. This could include an extension of the services provided by our internal laboratory or the Pump Test and Repair facility.

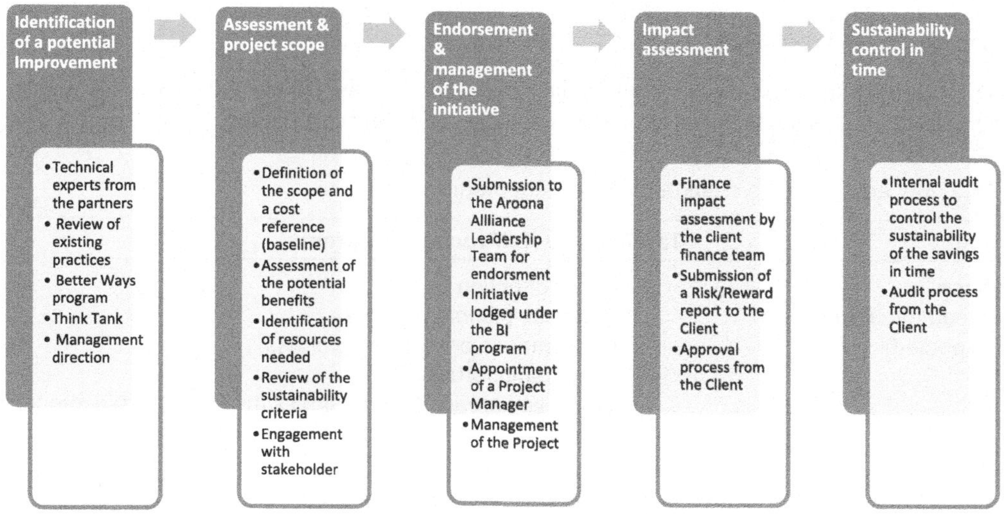

Figure 15.2 Process in place to implement business improvements and control the sustainability before and after the project being delivered.

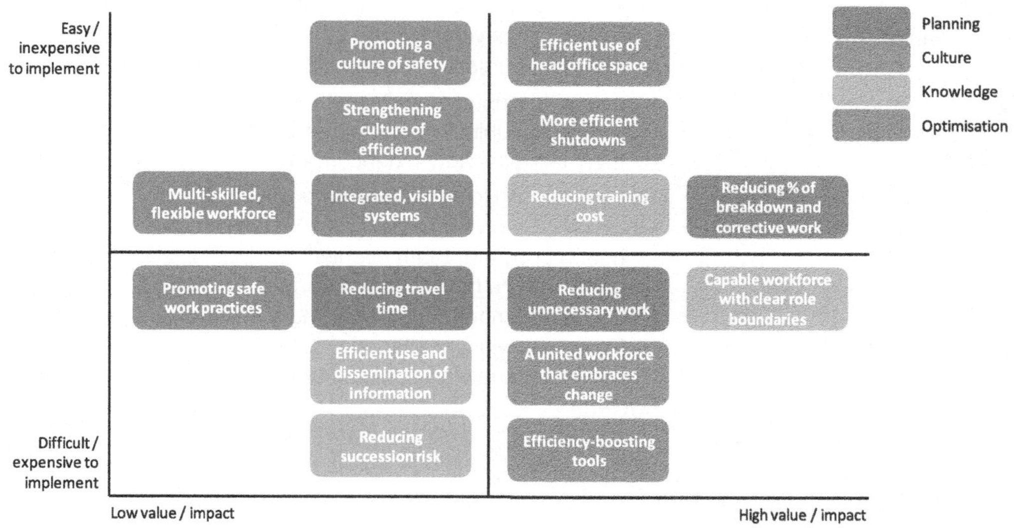

Figure 15.3 Key projects managed under the BI1 program.

15.1.7 The cultural and the change management challenge

As the Good to Great program has been designed to address all processes and practices in the business, a key critical success factor has been to adapt progressively our culture and manage proactively the change. A specific *change plan* has been designed and implemented by proposing strategies and tactics that:

- understand the employees and stakeholders participating in and impacted by the program;
- create energy and momentum to achieve the 2018 goals;
- effectively and efficiently move employees through the change cycle; and
- manage risk and resistance.

Table 15.2 illustrates what has been put in place to ensure there is sufficient awareness of why we need to change, desire to change, knowledge on how to change and ability to change.

Table 15.2 Key actions of the change management program.

Objective	Tactics – Tools/Process/Programs
Awareness (for the need for change and what's expected)	• Set clear direction – Business planning process – Performance agreement process – Leadership forums • Target all key areas of the business through multiple channels • Tools to support delivery of consistent messages
Desire (to be a part of the change)	• Broad communication • Coaching – one-to-one discussion on why • Leadership engagement – forums • Great place – giving back – health program, team events • Enable employee involvement – Better Way program, forums, think tanks, workshops, etc.
Knowledge (of how to change)	• Training for identified required skills • Leadership – development – specific courses, via leadership forum, etc.
Ability (to implement required skills and behaviours)	• Coaching to enhance skills in the workplace. • Exposure to subject matter experts and different thinkers – think tanks, reviews, benchmarking
Reinforcement (to keep us on track; make it stick)	• Recognition & Reward (corporate & local) • Great place – giving back – health program, team events – performance management (formal & informal) • Performance monitoring, reporting & communicating – Progress, wins, challenges, data, surveys • Consolidating new practices in the business (Quality Management System)

The impact of the change program has been measured through an *engagement survey* (done every 7 months across the Alliance). Since the inception of the program, survey results have improved (reaching 71 out of 100) and given important information on the areas of the business that needed more support, as well as checking that the key messages were understood and capable of driving our collective behaviour.

15.1.8 Main achievements two years after the Good to Great strategy inception

In June 2016, the Good to Great strategy reached 3 years of implementation. Significant results have been achieved to-date including almost 5% of cost savings across the 2015/2016 financial year, representing $16 million of cumulative sustainable savings since the inception of the Alliance, while achieving the contractual expectations.

Details of each **business improvement program** are given in Table 15.3.

Table 15.3 Sustainable cost efficiencies achieved and planned.

Business Improvements	FY 2012/2013 (k$) First Year of the Alliance	Good to Great Program (FY14 – FY18)			
		FY 2013/2014 (k$)	FY 2014/2015 (k$)	FY 2015/2016 (k$)	Targeted Over 5 Years
BI1 (O&M integrated model)	$392	$1,115	$1,584		$6,418
BI2 (Automation & control)	$0	$0	$0		$404
BI3 (Maintenance optimization)	$0	$663	$849		$3,262
BI6 (Process optimization)	$0	$163	$1,034		$2,573
BI7 (Energy saving & recovery)	$0	$175	$80		$2,148
BI8 (Procurement)	$547	$515	$341		$2,717
BI9 (Better ways)	$500	$234	$671		$2,194
BI10 (Head Office review)	$0	$408	$683		$3,378
BI11 (Additional services) & other external business improvements	$0	$69	$0		$1,907
Total sustainable savings Year (*n*) vs. Year (*n* – 1)	**$1,439**	**$3,341**	**$5,241**	**$4,801**	**$25,000**
Total cumulated sustainable savings	**$1,439**	**$6,219**	**$16,240**	**$31,062**	

15.2 CONCLUSION

The Water Corporation has set up the Aroona Integrated Alliance to make a step change in performance by combining the knowledge and expertise of its own staff with two private partner organizations, Suez and Broadspectrum.

After 4 years of the operation, Aroona has delivered a cumulative $31M in sustainable savings as well as significant improvements in the management of health and safety, and operational performance.

Critical success factors for this change were the capability to engage the workforce in this improvement program, and generate and keep the momentum – both from a short-term (Better Way program) and long-term perspective (Aroona Think Tank).

This innovative alliance structure has provided a framework and the governance model which demonstrates how to drive the right behaviour from all three partners, management and employees of Aroona.

Chapter 16

Experience from Eastern Europe

Augustin Boer, Teodor Popa and Dobromir Simidchiev

The utilities from Eastern European Countries were fusing in the last 10 years on aggregation of operation through regionalization and on the implementation of large scale investments mainly with EU and IFI financing. In this context, the measures for energy efficiency increase, decrease of NRW and increase of overall performance were not on top of the agenda of the management teams of the utilities. Because the tariffs have reached close to maximum affordable levels, there is an increase pressure from the authorities (shareholders) to start considering the efficiency increase measures as main driver for financial performance improvement. In this environment, the aim of this chapter is to highlight the existing situation in Eastern Europe; the possibilities, opportunities and challenges for having PBCs and present some basic study cases of projects that could be classified as micro-PBCs.

Performance-based contract (PBC) instruments are very limited in Eastern European countries, mainly as a result of sector focus on consolidation through regionalization and on large-scale investments preparation and implementation, financed mainly with European Union and IFI funds. Most of the operators from the region were focusing in the last 5–10 years mainly on increasing the connection rate to water and wastewater services and on assuring compliance with the environmental requirements (mainly for wastewater treatment), and less on performance improvement.

The aim of this chapter is to highlight the existing situation in Eastern Europe; the possibilities, opportunities and challenges for having PBCs and present some basic study cases of projects that could be classified as micro-PBCs.

The overall situation of the water sector in some of the countries from the region is the following:

- **Romania:** A total of 43 Regional Operating Companies (ROCs) have been formed through the consolidation of smaller operators into larger ROCs. They are providing water to around half of the country's population, and the percentage is increasing every year. Two private operators provide services in Bucharest and Ploiesti covering 9% of the population, while local municipal services and self-provision are the sources for the rest of the population.

Authors: Jan Janssens, Didier Carron, Philippe Marin, Philip Giantris and Tom Williams
doi: 10.2166/9781780405964_0155

- **Bulgaria:** Bulgaria has 64 water and wastewater operators, from which 29 regional companies are providing services to 76% of the population, the private operator in Sofia supplying services to 18% and municipal utilities and self-provision being available for the rest of the population. Although water coverage is almost entirely assured, the wastewater collection and treatment still need to be extended and improved. The consolidation (regionalization) of the sector is extensively encouraged and there are a few positive examples recorded in the last year.
- **Moldova:** In Moldova, the 52 municipally owned operators provide water and sanitation services to 43% of the population mainly in the urban areas, while the rest of the population relies on self-provision. The regionalization of water and wastewater services operation is a strategic objective included in the Water Supply and Wastewater Strategy of Republic of Moldova. In some of the Rayons, the process started by extending the area of operation of the utility companies from larger cities to the surrounding villages. The authorities from Moldova have under preparation or implementation with assistance from IFIs, investment projects considering larger scale regionalization of the water sector (e.g. the creation of a regional operator covering 6 Rayons and Balti Municipality in the north, regional operation in Straseni and Calarasi Rayons in the center, and regional operation in Cahul Rayon in the south, etc.).
- **Serbia:** In Serbia, 7 regional public utilities provide service to several municipalities covering almost 31% of the population, while 145 municipal public utilities serve 44% of the population, and the remainder relies on self-provision. There is a national strategy and policy framework for the water sector which is comprehensive and adequate, however, only very few elements from the strategy were actually implemented. There is a particular case regarding the economic regulation of tariffs: since 2004, the Ministry of Finance has imposed a ceiling on tariff increase and the public utility companies cannot exceed the programmed inflation rate.

The experience of the water utilities from the region regarding private operators or private partners is also very limited in the region with only a few cases being recorded in the last 20 years:

- **Romania:** The initiatives of involving the private sector in the water utilities management and operation were considered as a solution for the Romanian water sector in the late 90s leading to preparation of studies and signature of two contracts with private operators. The Municipalities of Bucharest and Ploiesti granted the operation of the water and wastewater services to Veolia for a period of 25 years in 2000. Some other studies were also prepared regarding private sector involvement (e.g. a concession study for Timisoara City in 1997, a concession study in Iasi County in 2005, etc.) but no other contract was signed due to opposition either of local or central authorities. After 2007 (the year of joining the European Union) the partnership idea with the private companies was stopped and the entire sector focused mainly on preparation and implementation of EU funded investments (the general message that circulated in the sector was that if private partners are involved, the EU funding might be significantly delayed or even lost).
- **Bulgaria:** The water and wastewater service in the capital city Sofia is delivered through a concession by a company jointly owned by Veolia and the Municipality. The contract was awarded in 2000 for 25 years. In recent years there has been a development of consultancy competences in providing O&M services to industrial customers related to industrial water and wastewater infrastructure.
- **Moldova:** There is no real participation of the private sector in Moldova. Two studies were made, one for the northern part of the country (Balti–Soroca pipeline) and the second for the

south. The intention was to find private sector opportunities in operating and extending the services, solving at the same time the existent supply and environmental problems. The project in Balti–Soroca is more advanced, the local authorities involved are taking the decision to be part of the regional operation and consolidate into a Regional Operator. The project envisages in future a tender for a private operator for a lease/affermage type contract. However, the project is progressing very slowly due to low involvement of central and local authorities in taking decisions related to the project. In 2013 Balti Municipality gave in concession the sewerage and wastewater activity to a private operator through tender. The concession however is not considered to be a success by the Municipality because the contractual balance was not assessed and implemented equitably, leading to serious problems for the remaining water publicly owned operator.
- **Serbia:** In Serbia, there has not been private sector participation in the water sector. Currently, the city of Belgrade is introducing a BOT project in the solid waste management sector. It is believed that the results of this project will influence the decision of the sector to look more toward the private operator's involvement.

In this context, the viability of PPP in general and PBCs in particular in the region was explored, mainly by EBRD. In 2014, a study was prepared in Romania with the goal to assess the possibility to improve the quality and coverage of water services and maximization of results by attracting private resources and alternative financial mechanisms. The study was discussed in detail with the entire sector but no follow up actions were recorded, mainly as result of the reluctance of the sector toward PPP.

In the second half of 2016, EBRD took the process a step forward in promoting the PBC instrument in Eastern Europe through two technical assistance projects:

- A Policy Dialogue Initiative that should provide interested governments in Bulgaria, Moldova, Romania and Serbia with information and guidance on the required reforms to facilitate private sector participation in PBC in the water sector. Under this project an international workshop was organized in Bucharest on September 2016 to discuss the PBC opportunities and constraints in the Eastern European context. As a result of this technical assistance, a policy paper on PBC is under preparation (as of June 2017, the document has not been issued officially).
- EBRD supports RAJA Constanta (the largest public regional operator from Romania) in introducing a PBC aimed at increasing operational efficiency and reduction of non-revenue water. The objective is to address the challenge in a systemic fashion, estimated at a fraction of the cost of pipe replacement, by dividing the city network into District Metered Areas and making punctual interventions to reduce water pressure and NRW. It is expected that the contractor will be partially remunerated based on its performance and achievements. If feasible, this could be the first actual PBC contract in the region. As of June 2017, the tender to prepare the PBC ToR has not been launched.

The main challenges to be overcome for having PBCs functioning in the region can be summarized as follows:

- Difficulties of ring fencing the private partner's activities: in the case of NRW issues, an obstacle is that NRW data for many companies from the region are not based on exact measurements and the area is not ring fenced due to a historical decision of the management team to report a lower level of losses owing to regulatory or fiscal reasons. Most of the companies from the region have under preparation or implementation, investment projects with EU financing, having NRW reduction components (e.g. network rehabilitation) which could overlap the area of the PBC, making it very difficult to ring fence and assess the "impact" of the private party measures and the "impact" of the EU or public funded investments.

- The legal and regulatory framework is a challenge for private sector involvement in the water sector. The typical problems regarding legal and regulatory framework are the following:
  - The regulatory body does not allow through the licencing requirements the handover of the system separate part operation to a private party. The regulations explicitly mention that any activity related to the supply of the service cannot be delegated to a private party.
  - In some countries, the national legislation is not very permissive regarding the tender of services for operating a part of the activity (which can be interpreted as subdelegation);
  - Due to lack of an explicitly approach in the existing legislation, a potential PBC contract needs to find exceptions or use interpretation of the legislations to be able to be implemented, which leads to lack of conformity and credibility of the decision makers;
- In some cases, the market is not sufficiently mature to accommodate PBC, owing to generally small-scale utilities and limited incentives for efficiency.

However, even in this somehow less favorable environment, there were in the past a few projects in the region (mainly pilot projects) that can be recorded as very small PBCs (or micro-PBCs).

One pilot project for energy efficiency was implemented by the Compania Apa Brasov (the regional operator from Brasov County in Romania). The project and the contract was designed in 2009 by a Danish financed technical assistance and consisted of the efficiency increase of a water pumping station from Brasov City. The initial idea was to have a PBC for energy consumption reduction at a large and important pumping station but, during the analysis and contract signing, a few problems were identified regarding the risk allocation:

- There was an important uncertainty regarding the future quantities of water that would be pumped. Until the date of the feasibility study, the quantities of water passing the pumping station were decreasing and, even if the forecast was considering a continuation of decrease, there was an important concern that the decrease would be more significant than envisaged. This uncertainty was mainly based on the high tariff increase that the company had to implement in Poiana Brasov (the area supplied by the pump to comply with the "polluter pays" principle which could trigger, due to the price elasticity effect, a high decrease in invoiced quantities. The sensitivity analysis showed that the decrease of quantity is a very sensitive variable for the future feasibility of the project and the private partner was reluctant to take over this risk;
- The utility wanted also additional rehabilitation investments to be performed at the pumping station (beside the energy efficiency ones) which did not assure in reasonable time the return of investment for the private partner;

Considering the elements mentioned above, in the end a contract with fixed payments was negotiated without having a performance-based component related payment.

The results of the energy efficiency project implementation and operation in Brasov utility are the following:

- The water quantities decreased significantly after the tariff strategy implementation, mainly due to the price elasticity effect (which triggered also a different consumption pattern and customer leakage reduction measures) recording a 29% decrease in 1.5 years.

- The increase in efficiency of the new pumps was recorded immediately and can be seen in Figure 16.1.
- The financial performances of the contract are presented in Figure 16.2.

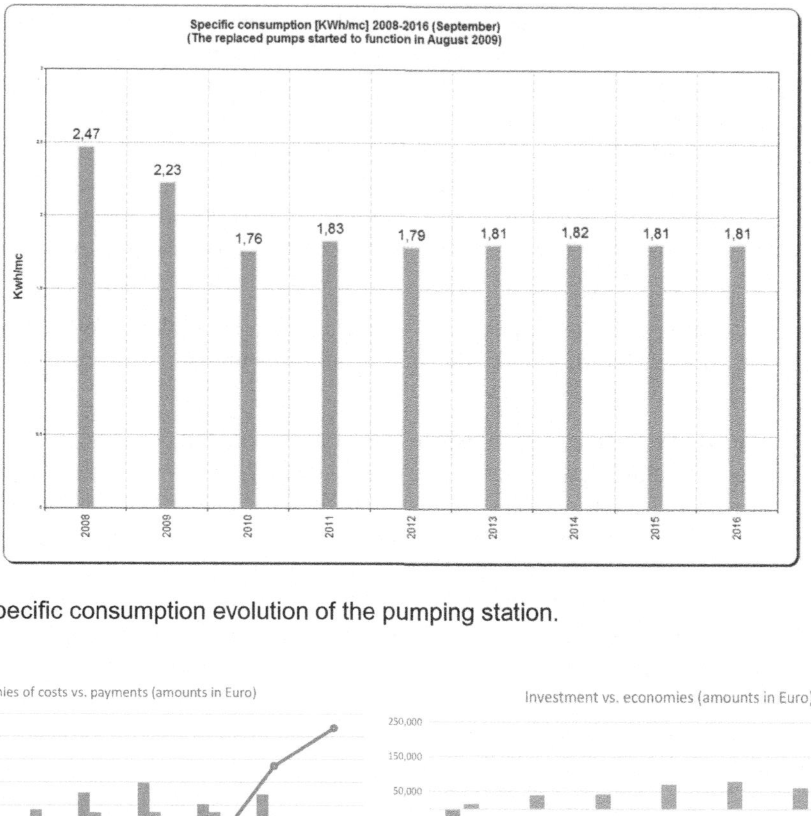

Figure 16.1 Specific consumption evolution of the pumping station.

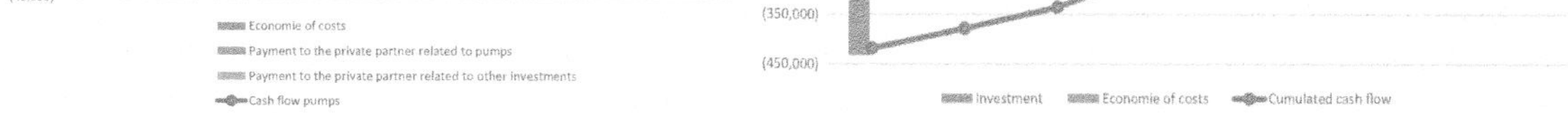

Figure 16.2 Financial performances of the project.

The investment in energy efficiency (the pump replacement investment) recorded very good results even with the decrease in quantity. The cash flow of this investment component assured a positive cash flow for the utility starting in 2013, generating also a cash surplus which was used to finance the payments for other investment components. This result was achieved also due to the increase in the market electricity price compared with the date of the feasibility study.

The overall investment will be fully recovered from the economies of electricity costs by the end of 2016.

The utility considers this project a success because they recorded economies of costs which assured over time the recovery of the investment. Also, during all this time, the pumps functioned without needing any repairs or replacement (only the standard maintenance activities were performed).

The second study case comes from Bulgaria. One of the biggest problems that some of the utilities are facing is the lack of operations expertise (including knowledge, human resources and technology), which is mainly critical in the case of relatively small and geographically scattered assets. This issue is valid both to municipal utilities and to industrial customers. Considering this environment, the company Hydrolia (a local private consulting company) decided to pool together the O&M activities of various asset owners within a centralized team that is heavily specialized in water management. Then through local presence, technology and 24/7 monitoring and communication the expertise is rolled out as daily operational service. In essence this means "having your distribution networks, treatment plants, reservoirs and pumping stations run as an outsourced service". Although the service is applicable for all aspects of WSS management, they decided to focus on core operations functions that require more sophisticated expertise such as:

- Treatment plants with particular emphasis on wastewater treatment plants (WWTPs);
- Distribution networks including zoning, dynamic monitoring and pressure control;
- Pumping stations and energy efficiency.

The activity developed was first focusing on industrial customers which need to have WWTP facilities, not having or willing to invest in high-level experts to supervise technological processes, sludge management, etc. The next step was to focus on municipal utilities. Small water operators (some of them as small as employing 15–20 staff) clearly cannot afford the luxury of having WWTP, SCADA, energy-efficiency or NRW experts as full-time employees. The company offers them the service of having such expertise on a remote, project or on-demand basis and they jointly decide how to integrate it in their daily operations.

For both categories of clients, the contracts have a performance based component. With industrial clients, the core focus is on reliability, environmental compliance and, not least, cost control. A very typical situation is when the customer is happy to pay the same level of OPEX in the long-term (and a fair amount of CAPEX) but to obtain an improved service. The implementation of the projects showed that the strengthening of the operational processes allows for both savings and improved parameters.

With municipal clients, the focus is shifted towards regulatory compliance and customer service. Typical key performance indicators include NRW (% of water lost as % of water produced), % of compliant samples at outlet of WWTP, energy used per cubic meter water produced, etc. The focus is also on the area of commercial and financial management of municipal water operators where common indicators are billed volumes per customer served, collection rates and cost recovery.

A relevant example of such shared services is a performance-based contract signed by Hydrolia with a small-size municipal operator providing public service to three middle-sized villages. The contract affects the operational and commercial performance of the operator and has a minor performance-based component in the remuneration (5%). The parameters to be improved are:

- Operational improvements
    - NRW at the inlet of the system:
        - Baseline level 70%
        - Target to be achieved 55% after 5 years.
    - Continuity of supply – baseline to be determined.
    - Failures of the water supply and sanitation network – baseline is set as global goal for both networks. The target for the water supply is to reduce the global number by half at the end of

year 3. The target for the sewer network is to remain at the current levels as the Capex is heavily restricted in it.
  - Water supply pressure – the target is to create at least six pressure management areas.
- Commercial improvements
  - Population coverage – the baseline is to be determined. The target is to achieve 98% by year 3.
  - Customers with revenue meter:
    - Baseline is 80%;
    - The target is 90% (from the increased customer database) by the end of year 3;
  - Invoiced water volumes target is to achieve 99% by the end of year 3.
  - Collection:
    - Baseline is 85%;
    - Target is to achieve 90% by the end of year 3.

Although the approach and the example presented is related to small-scale operation, for this region this can be a good starting example in implementing PBCs, which can be followed by the larger operators.

As a general conclusion, we can say that the water utilities from Eastern Europe started to look more and more toward efficiency improvements as the main strategic objective after 5–10 years of focusing mainly on large-scale investment implementation.

One of the biggest problems faced by the utilities from the region is related to non-revenue water and there are pressures from the authorities to improve the situation in the coming years. The energy consumption has and will continue to be increased significantly due to extension of operation in rural areas (areas with lower density of customers) and due to starting the operation of environmental compliant assets (wastewater treatment plants, etc.). Significant pressure is expected for future optimization of operation leading to increase of energy efficiency.

In this context, the PBCs can represent a feasible and viable solution which can accommodate the continuity of publicly owned company operation with the involvement of private experience for efficiency increase.

Chapter 17

Performance-based contracts (PBCs) for improving utilities efficiency: Experiences and perspectives: The way forward and perspectives/trends

Didier Carron and Jan Janssens

Performance-based contracts are definitely a valid option for contracting by water utilities but they need careful design to deliver on their promises.

17.1 SUMMARY OF DISCUSSIONS OF 31 OCTOBER 2016

1. PBCs in the water sector, not limited to NRW only, but also applicable and of interest in all water sector related domains.
2. PBCs justify calling upon third parties (often private) to achieve improved performance. A PBC however is not equal to privatization.
3. PBCs imply results-based remuneration; results can be either technical or financial.
4. There are two basic avenues:
   (a) Same result, but achieved at lower cost.
   (b) Improved result, with an optimized cost.
5. In essence there are two possible strategies:
   (a) As an operator: with focal attention on processes, improving performance by reducing costs (OPEX) using indicator ratios of operation (e.g. gains in volume of water through reducing NRW, which translate into reducing costs or in increasing revenues)
   (b) As a financier: with focal attention on results-based financing (the so-called *inverse approach*):
      - Option 1: available CAPEX is a constraint: how to achieve maximum possible performance (typically: highest possible network efficiency 1-NRW)
      - Option 2: target value of KPI (e.g. NRW) is a constraint: how to achieve this with minimum CAPEX (so-called *smart investment*, doing more with less)
6. Each strategy involves risks for the contracting parties. These risks need to be managed, shared and assumed. Parties will be outside of their comfort zone with uncertainties and potential surprises to

Authors: Jan Janssens, Didier Carron, Philippe Marin, Philip Giantris and Tom Williams
doi: 10.2166/9781780405964_0163

face over several years. The contract needs to be sufficiently flexible to allow necessary adaptations. The commitment of top management is essential for both parties.

7. Sustaining gains of performance after a PBC remains a challenge. Hence PBC would need to include a capacity building component. An additional year after a PBC to sustain performance gains is an interesting option. The Alliance Contracting model seems to emerge as a good practice.
8. The challenge is often defining gains of performance in monetary terms in domains other than NRW. A key principle would be to share financial results between principal and agent. The available budget however is often disconnected from results.
9. What definitely should be relinquished is an input-based engineering approach, as illustrated in Figure 17.1 below.

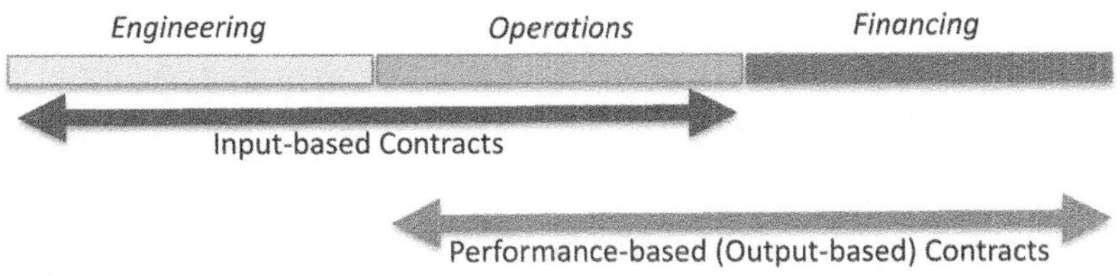

Figure 17.1 Scope of PBC applications.

10. Procurement of PBCs remains an issue. Model TORs and contracts that reflect the above strategies are not readily available.
11. There is a continuing need for capacity building of utilities. The erroneous perception that PBC is a hidden privatization still persists and is one of the reasons that utilities are often reticent. That a PBC is not a precursor for privatization is clearly demonstrated by the various case studies in the preceding chapters.

Index

A
Affermage contracts, 135–143
Alliance, 18, 91, 145–153

B
Billed volume, 53
Business improvement, 149–151

C
Change management, 152–153
Commercialization, 103
Continuity of supply, 128, 131, 160
Contract, 53, 56, 58, 129, 164
Contracts, 129, 135–143
Corporate commitment, 113
Cost efficiency, 48, 149

D
Design, 2, 19, 91–97, 124–127

E
Earning logic, 138–139
Eastern Europe, 155–161
Economic leakage level, 12
Efficiency, 4, 10
Enabling environment, 115
Energy conservation, 15, 42
Energy efficiency, 15–22, 159
Energy saving companies, 3

F
Financial analysis, 45, 49
Financing, 2, 69

G
Governance, 90–91, 147–149

I
Incentive system, 114
Incentive, 27, 75, 147
Inverse approach, 7–13

K
Knowledge transfer/capacity building, 91
KPI, 1, 2, 4, 19, 21, 31, 89–91, 93–95, 100, 104, 107, 111, 113, 117, 121, 124, 126–129, 131, 133, 134, 142, 143, 163

L
Leakage, 67, 70, 82

M
Marginal cost, 75
Micro-PBC, 155, 158
Minimum service level, 104, 108
Municipal, 34, 156, 160

N
National Water and Sewerage Corporation reform, 23, 26
Navi Mumbai, 33–42
New public management, 23
Non-revenue water management, 137
Non-revenue water (NRW), 4, 5, 7, 8, 17, 21, 25, 45, 54, 63, 75, 91, 126, 127

O
Optimization, 75–88
Outcomes, 63, 66

Output-based contracts, 37
Outputs, 63, 66

P
Performance-based contracts, 1–6, 15–22, 39, 45, 53, 63–73, 89, 103–108, 116, 121–134, 145–153, 155, 163–164
Performance contracts, 23, 24
Performance improvement, 27, 31, 114
Performance monitoring, 40–41
Performance targets, 103, 114, 126
Performance-based remuneration, 55, 127
Private sector participation, 26, 33, 42
Procurement, 40, 59, 72–73, 127–129
Profit sharing, 135
Progressive realization, 104
Public enterprise management, 23, 26
Public–private partnership (PPP), 1–2, 38, 40, 89, 90, 101, 110

R
Remuneration, 125, 131–132
Results-based financing, 1, 7–13
Right to water, 104
Risk allocation, 158–159

S
Sanctions, 108
Sanitation, 31, 123
Service continuity, 57, 93
Sustainable network management, 70
Sustainable results, 67

T
Targets, 29, 58, 60, 68, 75, 92–93, 140
Tariff, 26, 38, 79, 96, 136, 155–156
Tax, 98, 136
Tegucigalpa, 53–62

U
Uganda, 23–31
USAID-FIRE (D), 33
Utility turnaround, 109, 112

V
Value for money, 139, 146, 148

W
Water operator partnership, 110–113

IWA Publishing's authorised EU representative for General Product Safety Regulations is Diane D'Arras, 15 rue Duret, 75116 Paris, France, e-mail: safety@iwap.co.uk.

Printed and bound by CPI Group (UK) Ltd, Croydon, CR0 4YY
16/06/2026
02140380-0003